[英国] 迈克尔·霍斯金 著　陈道汉 译

牛津通识读本·

# 天文学简史

# The History of Astronomy

## A Very Short Introduction

译林出版社

图书在版编目（CIP）数据

天文学简史／（英）霍斯金（Hoskin, M.）著；陈道汉译. —南京：译林出版社，2013.5（2023.10重印）
（牛津通识读本）
书名原文：The History of Astronomy: A Very Short Introduction
ISBN 978-7-5447-2981-9

Ⅰ.①天… Ⅱ.①霍… ②陈… Ⅲ.①天文学史－世界
Ⅳ.①P1-091

中国版本图书馆 CIP 数据核字（2012）第 133146 号

著作权合同登记号　图字：10-2014-197 号


**天文学简史** [英国] 迈克尔·霍斯金 ／ 著　陈道汉 ／ 译

责任编辑　　於　梅
责任印制　　董　虎

原文出版　Oxford University Press, 2003
出版发行　译林出版社
地　　址　南京市湖南路 1 号 A 楼
邮　　箱　yilin@yilin.com
网　　址　www.yilin.com
市场热线　025-86633278
排　　版　南京展望文化发展有限公司
印　　刷　江苏苏中印刷有限公司
开　　本　890 毫米 × 1260 毫米　1/32
印　　张　7.625
插　　页　4
版　　次　2013 年 5 月第 1 版
印　　次　2023 年 10 月第 13 次印刷
书　　号　ISBN 978-7-5447-2981-9
定　　价　39.00 元

# 序言

江晓原

　　天文学作为一门自然科学，有着与其他学科非常不同的特点。例如，它的历史是如此悠久，以至于它完全可以被视为现今自然科学诸学科中的大哥（至少就年龄而言是如此）。又如，它又是在古代世界中唯一能够体现现代科学研究方法的学科。再如，它一直具有很强的观赏性，所以经常能够成为业余爱好者的最爱和首选；而其他许多学科——比如数学、物理、化学、地质等等——就缺乏类似的观赏性。

　　由于天文学的上述特点，天文学的历史也就比其他学科的历史具有更多的趣味性，所以相比别的学科，许多天文学书籍中会有更多令人津津乐道的故事。例如，法国著名天文学家弗拉马利翁的名著《大众天文学》里面充满了天文学史上的遗闻轶事——事实上，此书几乎可以当做天文学史的替代读物。

　　西人撰写的世界天文学通史性质的著作，被译介到中国来的相当少，据我所知此前只有三部。这三部中最重要的那部恰恰与本书大有渊源——那就是由本书作者霍斯金主编、被西方学者誉为"天文学史唯一权威的插图指南"的《剑桥插图天文学史》（*The Cambridge Illustrated History of Astronomy*）。

　　霍斯金（Michael Hoskin）是剑桥丘吉尔学院的研究员。退休前曾在剑桥为研究生讲授天文学史 30 年。在此期间他还曾担任科学史系系主任。1970 年他创办了后来成为权威刊物的《天文学史》杂志(*Journal for the History of Astronomy*)并任主编。在国际天文学联合会(International Astronomical Union)和国际科学史与科学哲学联合会(International Union for the History and Philosophy of Science)的共同赞助下,他还担任由剑桥大学出版社出版的多卷本《天文学通史》(*General History of Astronomy*)的总主编。而这本《天文学简史》则可以视为上述多卷本《天文学通史》的一个纲要。

天文学的历史非常丰富,但是在传统观念支配下撰写的天文学史,则总是倾向于"过滤"掉许多历史事件、人物和观念,"过滤"掉人们探索的过程,"过滤"掉人们在探索过程中所走的弯路,"过滤"掉失败,"过滤"掉科学家之间的钩心斗角……最终只留下一张"成就清单"。通常越是篇幅较小的通史著作,这种"过滤"就越严重,留下的"成就清单"也越简要。本书正是这样一部典型作品。

这种作品的好处是,读者阅读其书可以比较省力地获得天文学历史发展的大体脉络,知道那些在传统观念中最重要的成就、人物、著作、仪器、方法等等。这类图书简明扼要,读后立竿见影,很快有所收获。

这种作品的缺点是,读者阅读其书所获得的历史图景必然有很大缺失——归根结底一切历史图景都是人为建构的,故历史哲学家有"一切历史都是思想史"、"一切历史都是当代史"这样的名言。人为建构的历史图景,永远与"真实的历史"——我们可以假定它确实存在过——有着无法消除的距离。

历史图景之所以只能是人为建构的,根本原因之一就在于史料信息的缺失。而历史的撰写者,无论他撰写的史书是如何卷帙浩繁、巨细靡遗,都不可能完全避免上面谈到的"过滤",这就进一步加剧了史料信息的缺失。况且每一个撰写者的过滤又必然不同,结果是每一次不同的过滤都会指向一幅不同的历史图景。

所以,历史永远是言人人殊的。

2010 年 3 月 25 日
于上海交通大学科学史系

# 目录

第一章
# 史前的天空

　　天文史学家主要依靠遗存下来的文献(古代文献在数量上比较零散,占压倒性多数的文献出自近代)以及仪器和天文台之类的人造物进行研究。但是,在文字发明**之前**就生活于欧洲和中东的人的"宇宙观"中,我们能够发现天空所起的某些作用吗?是否曾经甚至有过一种史前的天文科学,使得当时的某个杰出人物得以预告交食现象?

　　为回答这些问题,我们主要依仗遗存下来的石碑——它们的排列、它们和地形的关系以及我们在某些石碑上发现的雕刻(通常是意义不明的)。当某一块石碑很独特时,根本的判断方法问题就最容易受到争议。例如,巨石阵在一个方向朝向夏至日的日升,而在另一个方向则朝向冬至日的日落。我们怎么能认定,这样一种在我们看来具有天文学意义的排列,正是巨石阵的建筑师因为该理由而选择的呢?它会不会是出于某种非常不同的动机或甚至是纯属偶然呢?另举一个例子,一座建于公元前3000年左右的石碑朝向东方,可能是因为金牛座中的亮星团即昴星团在东方升起,可能是因为它朝向夏至和冬至日升方向的中点,可能是因为在那个方向有一座神圣的山,或者选择这个方向只不过是为了利用地面的坡度。我们如何能判定,

1

建造者怀有的是其中的哪一个想法(如果有的话)?

当论及散布于广阔地域中的大量石碑时,我们就不会那么盲目了。西欧的考古学家研究了石器时代晚期(新石器时代)的公墓,那个时候狩猎者的游牧生活已经被农夫的定居生活所取代。这样的坟墓为氏族的需要服务了许多年,因而它们都有一个入口,当有需要时,其他的尸体可由此放入。我们能够确定,坟墓的朝向正是里面的尸体通过入口向外"眺望"时的视线方向。

天文学简史

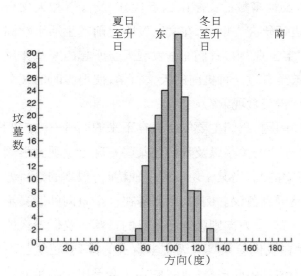

图1 葡萄牙中心区及邻近的西班牙地区177座七石室坟墓朝向的直方图。当计入地平高度后,我们发现,每座坟墓都在一年的某个时候朝向太阳升起的方向,大多是在秋天的月份里,我们可以料想当时建造者正好有闲从事这样的工作。这一点符合将坟墓朝向开工这天太阳升起方向的习惯,如同后来在英格兰和别的地方建造基督教堂时所惯常实行的那样。

在葡萄牙中心区有很多这样的坟墓，它们具有独特的而且是瞬间即可辨认的形状和构造，由习俗相同的人们所建造。它们散布在东西长约二百公里，南北宽度也近二百公里的一个无山地区，但是作者曾经测量过的177座坟墓全都面朝东方，在太阳升起的范围以内。

不仅如此，秋冬季节太阳升起的方向也是被优先考虑的。现在我们从书面记载得知，在许多国家基督教堂的传统朝向为日升方向（一年中两次），这是因为冉冉升起的太阳是基督的象征；建造者通常在建设开始之日使教堂面向日升方向来保证这一点。假定新石器时代的这些坟墓建造者遵循相似的习俗，假设他们也将升起的太阳视为一个生命来临的象征，那么既然他们无疑在收获之后的秋冬季节才有空闲从事诸如此类的工作，于是我们就有望发现我们实际所发现的朝向模式；难以想象任何其他解释可以说明这种引人注目的朝向。所以，推断新石器时代的建造者将他们的坟墓朝向定为日升方向，应该是合理的。

如果事实确是如此，则我们有证据认为，天空在新石器时代宇宙观中所起的作用，就同它在教堂建造者的宇宙观中曾起过（和正起着）的作用一样，但是，这同"科学"无关。主张史前欧洲的确存在一种真正的天文学的是几十年以前的一位退休工程师亚历山大·汤姆，他查勘了英国境内的几百个石圈①。汤姆认为，史前的建造者在设置石圈的位置时确保从这些位置看出去，太阳（或月亮）会在某个重要的日子——例如，就太阳而言，

---

① 巨石构成的环形遗迹。——书中注释均由译者所加，以下不再一一注明。

为冬至日——在一座远山的背后升起（或落下）。在至日前后几天以内，太阳差不多在地平线的同一位置升起（或落下），只有用很精密的仪器，才可以确定至日的正确日期。依据汤姆的说法，史前的杰出精英们利用石圈和远山构成了范围达方圆好多英里的仪器；他们利用太阳周和太阴周的知识，能够预报交食现象并由此确立了他们在人群中的优势地位。

汤姆的工作激起了人们巨大的兴趣，当然也引发了争议。但是，人们重新调查他的研究处所后能得出这样的结论：他知道他挑出的那些远山会符合其观念，而这样的排列可能纯属偶然并且和史前建造者没有任何关系。现在几乎没有人相信汤姆的猜想了，虽然任何一个试图理解史前宇宙观的人都应该因为他将注意力引向这样的问题而感激他。

我们可以肯定，在史前时期，天空至少为两类人（航海者和农夫）的实际需要服务。今天，在太平洋和别的地方，航海者利用太阳和恒星探寻他们的航程。史前地中海的水手无疑也是如此，但是在这方面几乎没有什么资料留存下来。关于农历——农夫始终需要知道何时播种及何时收获——我们倒有些线索。即使在今天，在欧洲有些地方，农夫还在利用希腊诗人赫西奥德（约公元前 8 世纪）在《工作与时日》中为我们描述的天体信号类型。每年太阳在恒星之间完成一次巡回，所以某颗恒星（例如天狼星）会因为太靠近太阳而有几个星期在白昼不可见。但是，随着太阳的继续运动，天狼星在拂晓的天空中闪现的日子就会来临，这一刻即为"偕日升"。赫西奥德描述了偕日升序列，他那时的农夫把这一序列用于他们的历法中，而这就定然将前几个世纪里汇集起来的知识和经验浓缩纳入其中。令人惊讶的是，似乎有早得多的这样一个序列被铭刻在马耳他姆那德拉寺院的柱子上,这个寺院可追溯到公元前 3000 年左右。我和我的同事找到了一连串似为计数单位的雕刻的小洞,在分析了数目之后,我们发现它们很好地表述着一次重要的偕日升和下一次之间所隔的日数。正如我们将要看到的,天狼星的偕日升很快就在附近埃及的历书中起到了关键的作用。

第二章
# 古代天文学

现代天文学的开端最初在公元前第三个和第二个千年的史前迷雾中浮现，起始于在埃及和巴比伦发展起来的日趋复杂的文化。在埃及，一个辽阔王国的有效管理依赖于一部得到认可的历法，而宗教仪式要求有在夜间获知时刻以及按基本方向定出纪念物（金字塔）方位的能力。在巴比伦，王位和国家的安全依赖于正确解读征兆，包括那些在天空中被见到的征兆。

因为在太阴月或太阳年中没有精确的日数，同样在一年中也没有精确的月数，所以历法历来是，现在也依然是难以制定的。我们自己月长度的异常杂乱正说明这是自然界向历法制定者提出的一大难题。在埃及，生活为一年一度的尼罗河泛滥所主宰。当人们注意到这种泛滥总是发生在天狼星偕日升前后，也就是当这颗恒星在经历几周的隐匿后再度出现于破晓的天空中时，他们就找到了历法问题的一种解决方案。因此，这颗恒星的升起可以被用来制定历法。

每年由 12 个朔望月和大约 11 天构成，埃及人从而制定出一种历法，其中天狼星**永远**在第 12 个月中升起。倘若在任一年中，天狼星在第 12 个月中升起得早，来年就还会在第 12 个月中升起；但若在第 12 个月中升起得晚，则除非采取措施，否则来年天狼星将在第 12 个月过完之后才升起。为了避免这样的事发生，人们就宣布本年有一个额外的或"插入的"月。

这样一种历法对于宗教节庆而言是适宜的，但对于一个复杂的和高度组织化的社会的管理而言则不然。所以，为了民用目的，人们制定了第二种历法。它非常简单，每年都是精确的 12 个月，每个月由 3 个 10 天的"星期"组成。在每年的末尾，人们加上额外的 5 天，使得一年的总日数为 365 天。因为这种季节年实际上稍长数小时（这就是为什么我们有闰年），所以该行政历法按照季节缓慢地周而复始，但是为了管理上的方便而采用这样一种不变的模式还是值得的。

因为有 36 个 10 天构成的"星期"，所以人们在天空中选用 36 个星群或"旬星"使得每 10 天左右有一颗新的"旬星"偕日升起。当黄昏在任一夜晚降临时，许多旬星将在头顶显现；到了夜晚，地平线上将每隔一段时间出现一颗新的旬星，标志着时间的流逝。

天空在埃及的宗教中起着重要的作用，因为在其中神祇以星座的形式出现，埃及人在地球上花费了巨大的人力，以保证统治着他们的法老有朝一日会位列其中。公元前第三个千年，法老的殡葬金字塔几乎精确地按南北方向排列成行，我们从中看到了一些端倪，至于这一排列是如何实现的，已有诸多争论。一个线索来自排列的微小误差，因为这些误差随建造日期而有规律地变化。最近有人提出，埃及人有可能是参照一条虚拟的线，这条线连接两颗特殊的恒星。在所有时间里，这两颗恒星都可以在地平线上见到（拱极星），当该线垂直时，就取朝向这条线的方向为正北。如其如此，由于地轴摆动（称为进动）所致的天北极的缓慢运动就可以解释这种有规律的误差。

埃及人为他们几何学和算术上的原始状态所制约,对恒星和行星的更难以捉摸的运动不甚了了,尤其是他们的算术几乎是无一例外地用分子为 1 的分式来运算。

相比之下,在公元前两千年,巴比伦人发展了一套算术符号,这一项了不起的技术成为他们在天文学上获得显著成就的基础。巴比伦的抄写员取用一种手掌大小的软泥版,在上面用铁笔刃口向外侧刻印表示 1,平直地刻印表示 10,按需要多次刻印,他就可以写出代表从 1 到 59 的数字,但是到 60 时,他就要再次使用 1 的符号,就像我们表示 10 这个数字那样,类似地可以表示 $60 \times 60, 60 \times 60 \times 60$,等等。在这个 60 进制的计数系统中,可以书写的数字的精确性和用途是没有限定的,甚至在今天我们仍然在继续使用 60 进制来书写角度以及用时、分和秒来计算时间。

巴比伦宫廷官员对所有种类的征兆都保持着警惕——尤为关注的是绵羊的内脏——他们保留着任何一个不受欢迎但接着发生的事件的记录,以便从中吸取经验:当征兆在未来再次发生时,他们就会知道即将到来的灾难的性质(该征兆是一种警示),于是就可以举行适当的宗教仪式。这就促使人们汇编了一部包含 7000 个征兆的巨著,它成形于公元前 900 年。

此后不久,为了使他们的预测更为精确,抄写员开始系统地记录天文(及流星)现象。这样的记录延续了 7 个世纪,太阳、月球和行星运动的周期开始逐渐地从记录中显露。借助于 60 进制计数法,抄写员设计出运算方法,利用这些周期来预报天体的未来位置。例如,太阳相对于背景星的运动在半年中加速,在半年中减速。为了模拟这种运动,巴比伦人设计了两个方案:或者假定半年采用一个均匀速度,半年采用另一个均匀速度;

或者假定半年采取匀加速运动,半年采取匀减速运动。两者都仅仅是对真实情况的人为模拟而已，但他们完成了这项工作。

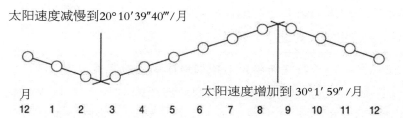

太阳速度减慢到20° 10′ 39″ 40‴ / 月

月

太阳速度增加到 30° 1′ 59″ / 月

12  1  2  3  4  5  6  7  8  9  10  11  12

图 2　巴比伦人对太阳相对于背景星运动速度的第二套模拟方案的现代表达。其中的数值见于公元前 133 或前 132 年的泥版。在这个人为的但却便于运算的表示法中，太阳的速度被想象成在 6 个月中作匀加速运动，然后在随后的 6 个月中作匀减速运动。人们发现这一方案的准确性能令人满意。

对于公元前 4 世纪以前的希腊天文学,我们的知识非常零碎,因为很少有那个时期的记载留传下来,而我们所拥有的,很多是即将被亚里士多德(公元前 384—前 322)抨击的主张中的引证。但有两个方面引起了我们的注意:首先,人们开始完全按自然的条件来理解自然,而没有求助于超自然;第二,人们认出了地球是个球状体。亚里士多德正确地指出,月食时地球投射

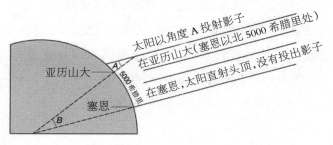

太阳以角度 A 投射影子

在亚历山大(塞恩以北 5000 希腊里处)

亚历山大

在塞恩，太阳直射头顶，没有投出影子

塞恩

图 3　埃拉托斯特尼为了测量地球所用的几何学,角度 A 和 B 是相等的。

在月面上的影子总是圆的,只有当地球是一个球体时,才会如此。

希腊人不仅知道地球的形状,而且埃拉托斯特尼(约公元前276—约前195)还对地球的实际大小做出了相当准确的估计。从那以后,受过少量教育的人都知道地球是球形的。

---

**埃拉托斯特尼对球形地球周线的测量**

埃拉托斯特尼相信,在现称阿斯旺①的地方,夏至日的正午,太阳位于头顶,而在阿斯旺正北5000希腊里的亚历山大,太阳的位置与太阳直射头顶处的距离为一个圆周的五十分之一。如其如此,则简单的几何学显示,地球的周长是5000希腊里的50倍。希腊里的现代等值是有争议的,但无疑250000希腊里这个值是近似正确的。

---

看起来,天空也是如此。而且,我们始终看见的正好是天球的一半,因此地球必定是位于天球的中央。于是经典的希腊宇宙模型形成了:一个球形地球位于一个球形宇宙的中心。

在艾萨克·牛顿时代仍被用于剑桥大学教学的亚里士多德多卷著作中,亚里士多德比较了位于宇宙中心的地球区——几乎延伸到远至月球处——和位于其外的天区。在地球区,变化、生死、存灭都在发生。地球在最中心;环绕着地球的是水层,然后是大气层,最后是火层。物体由这些要素按不同比例构成。在没有外力的情况下,物体会按直线运动,或者向着地心或者背离地心,从而使得离地心的距离合乎其元素构成。于是,本质似泥土的石头向着地心坠落,而火焰则向着火球升腾。

---

① 古称塞恩。

紧接着,火球之外就是天区的开始。在天区里,运动是周期性的(从不是直线运动),所以不存在真正的改变。天空最高处是转动着的球层,由不可计数的"固定"恒星构成,之所以说"固定",是因为恒星的相对位置从不改变。不固定星体的数目正好是 7 个:月球(明显是所有星体中最近的)、太阳、水星、金星、火星、木星和土星。这些星体相对于固定恒星运动着,并且因为它们的运动是永远变化的 (的确,5 个较小的星体实际不时地反向运动),所以它们被通称为"流浪之星"或"行星"。亚里士多德的老师柏拉图(公元前 427—前 348/ 前 347)天生是个数学家,曾视行星为对他的信念(我们生活在一个受规律支配的和谐宇宙中)的一种可能的反驳。但是,这是否也可能表明,行星的运动实际上像恒星的运动一样具有规律,唯一的差别是支配行星运动的规律更为复杂而不是一目了然?

应对挑战的是几何学家欧多克斯 (约公元前 400—约前 347),他为每颗行星设定了一个由三四个同心球构成的叠套系统,用于以数学方法演示行星的运动毕竟是似有规律的。他想象每颗行星位于最内球体的赤道上,该球作匀速转动并携带着行星运动,它的极被认为嵌入边上的球中并被其带动着也作匀速转动。第三个和(就较小行星而言)第四个球的情况也是如此。每个球的转动轴的角度都经过仔细地选择,其转动的速度也是如此。每种情形下,最外层的球产生该行星绕地球的周日轨迹,例如,月球诸球分别按 24 小时、18.6 年和 27.2 日的周期作匀速转动,所以月球的合成运动反映着所有这三个周期。

对于 5 颗较小行星中的每一颗而言,其中两个天球的速率相等但方向相反地绕着差别细微的轴转动,因而这些球本身将赋予行星一种呈现 8 字形的运动,使得由四球叠套而成的整个

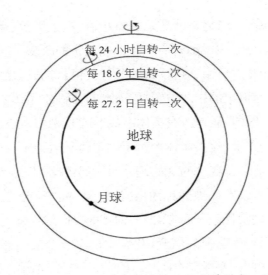

每 24 小时自转一次

每 18.6 年自转一次

每 27.2 日自转一次

地球

月球

图 4 依照欧多克斯的说法,月球运动所展示的数学模式。设想月球位于最内球层的赤道上,该球层每月旋转一周,这个球的极嵌入中层天球,中层天球每 18.6 年旋转一周,这个周期与交食周期相似,中层天球的极嵌入外层天球,外层天球每天旋转一周。

系统不时地产生向后的运动。

至此,一切是如此美好。但在这些几何模型中,较小行星的向后运动(逆行)完全有规律地重复着并且明显不重现那种在天空中我们实际所见的行星的乖僻运动。而且,模型迫使每颗行星维持在与中心地球距离恒定的位置上,而在真实世界里,较小行星的亮度以及与我们的距离变化都很大。这样的缺点会使一个巴比伦人觉得讨厌,但该模型足以满足柏拉图那代人,他们认为宇宙确实是有规律的,即使其规律有待于被完全阐明。

亚里士多德经受着非常不同的限制:模型中的球是数学家心目中的构建物,从而没有用物理术语说明,行星如何如我们所见的那样开始运动。他的解决办法是将数学的球转换成物理

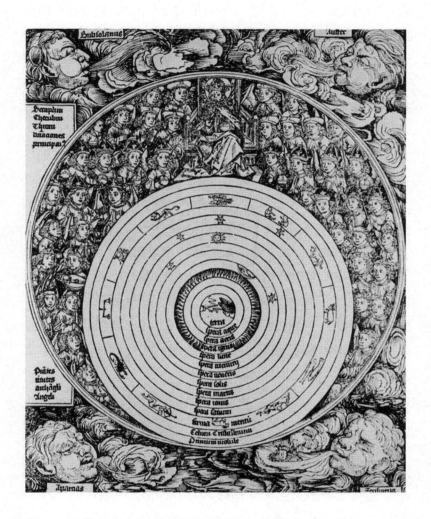

图 5　1493 年出版的《纽伦堡编年史》描绘了基督教化了的亚里士多德的宇宙。中心是 4 种元素(土、水、气、火),然后是行星球(月球、水星、金星、太阳、火星、木星和土星),再接着是恒星天球、水晶天堂球和第一推动者。最外面我们看到的是上帝被 9 个等级的天使簇拥着。

实体，并将它们联合起来为整个系统构建一个组合的叠套，所有球层的最外一个，也就是固定恒星的那个天球，其周日旋转可以对位于其内的每颗行星施加一个周日转动，所以每颗行星叠套球层的最外一个球可以舍弃。但是，除非采取步骤防止，专为个别行星设计的球会把它们的运动传递给系统，因此亚里士多德在合适的地方插入了反向转动的球层，旨在抵消不需要的转动。

　　所得到的亚里士多德的宇宙学——一个位于中央的地球或月下区，在那里有生存和消亡，以及一个位于其外的天区，其天球产生恒星和行星的周期运动——在 2000 年的大半时间里，支配着希腊人、阿拉伯人和拉丁人的思想。然而，在亚里士多德的门生亚历山大大帝征服了已知世界的许多地方，从而使希腊的几何天文学开始融入巴比伦人的算术天文学和观测天文学之后，希腊几何天文学的不具变通性以及所得到的理论和观测之间的差异几乎是立即获得了修正。匀速圆周运动继续被希腊天文学家视为理解宇宙之钥，但他们现在拥有了更多的变通性并且更多地关注观测事实。

图 6　在一个偏心圆上，行星照例绕地球作匀速圆周运动，但是因为地球不在圆心位置，所以行星的速度从地球上看起来会有变化。

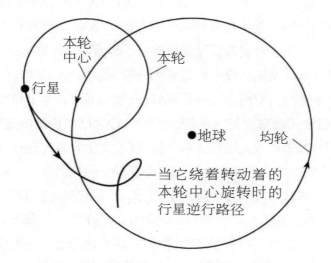

図 7　本轮是一个小圆,它携带着行星沿着这小圆作匀速运动。本轮的中心同样以匀速绕地球在一个均轮上运动。该图显示,只要适当选择两种速率,不时观测到的 5 颗较小行星的逆行运动就可以用这个模型来模拟。

公元前 200 年左右,几何学家佩尔加的阿波罗尼乌斯发展了两种几何方法来提供变通性,一种是把行星绕地球的运动视为匀速圆周运动,但该圆周相对于地球而言是**偏心**的。

结果是当行星的轨道较靠近地球时,行星看起来就会运动得快一些。当行星在远离地球的一边时,运动就会显得缓慢。在另一种方法中,行星位于一个小圆即**本轮**上,而本轮的中心则在一个**均轮**上绕地球转动。

我们很容易评价这一设计的价值。因为,正如我们所见,金星(或别的星体)环绕着太阳转动,而太阳又环绕着地球转动。人们也许会说,天文学在正确的轨道上,不只是在正确的轨道上,而且是在一条最有希望的轨道上,因为重复地修正所涉及的各种量(参数)会引起鼓舞人心的进展,但却从来没有获得过完全的成功——直到最后开普勒放弃了圆而采用椭圆为止。

第一个使用这些设计的是喜帕恰斯,他在公元前 141 年和前 127 年之间在罗得岛进行了观测。虽然他只有一本著作留传下来,随后托勒密将其纳入《天文学大成》时已变得陈旧过时,但我们获悉他的成就还是依赖于阅读《天文学大成》。通过喜帕恰斯,希腊人的几何天文学开始整合在漫长的几个世纪中所导出的精密参数。在那几个世纪里,巴比伦人保存着他们的观测记录。喜帕恰斯汇编了一部自公元前 8 世纪在巴比伦观测到的月食总表,这些记录对他研究太阳和月球的运动起了决定性的作用,因为在交食时,这两个天体和地球正好排成一线。喜帕恰斯采用了巴比伦的 60 进制书写数字,并且将黄道圈和其他的圆划分为 360 度。他只用一个偏心圆就成功地还原了太阳的运动。托勒密几乎原封不动地接受了这个模型。他在研究月球运动方面不怎么成功,而将较小行星的运动留给了他的后继者。

喜帕恰斯独一无二的最重要发现是分点的岁差,即黄道与赤道相交而成的两个相反位置在众恒星之间的缓慢移动。春分点被天文学家用于确定他们的参数坐标系,而该点的移动意味着恒星的测量位置随测量日期而变化。

喜帕恰斯也编制了一部恒星星表,但是业已佚失;唯一由古代留传下来的星表是《天文学大成》中的那一部。历史学家就此展开了辩论:托勒密是自己观测到他在星表中给出的位置,还是采用了喜帕恰斯的观测位置,在作了岁差改正后,简单地将恒星位置转化成他自己的历元?

在喜帕恰斯和托勒密之间的 3 个世纪,是天文学的黑暗年代。至少，托勒密似乎轻视那一时期的成就，也很少述及。大多数信息我们是从后来桑斯克里特的著作中获知的，因为印度天文学史很保守，而且它的作者保存着他们从希腊人那里所学到的知识。但是，《天文学大成》一书本身是比较稳妥的。关于该书作者的生平，我们所知寥寥，但是作者报告了他在127年和141年之间在伟大的文化中心亚历山大所作的观测，故而他的出生不可能会比2世纪的开始晚得太多。他可能是在博物馆和图书馆的发源地亚历山大度过了他的成年时期，并且他和喜帕恰斯一样，是在远离希腊大陆而又接近巴比伦人不可替代的观测记录点活跃的一位希腊天文学家。

《天文学大成》是一部权威著作，书中给出了几何模型和有关的表格，可以被用来计算在无限期的未来时刻太阳、月球和5颗较小行星的运动。该书写于亚里士多德之后500年，当时希腊文明近乎自然地发展着，书中综合了希腊和巴比伦在掌握行星的运动方面的成就。它的星表包含了被编排成48个星座的超过1000颗恒星，给出了每颗恒星的经度、纬度和视亮度。早先作者（著名的有喜帕恰斯）的著作因为业已陈腐过时，所以从地球上消失了，而《天文学大成》则像巨人一样在随后的14个世纪中统治了天文学。

但是，后来又出现了问题。亚里士多德的宇宙论用以地球为中心的同心球来阐释天空，哲学家对这样的球以及它们的匀速转动感觉良好。然而，阿波罗尼乌斯和喜帕恰斯则引入了破坏这一常规习俗的偏心圆和本轮。在这样的模型中行星的确还在圆上作匀速转动，但是并非以地球为圆心。这已经是够糟糕的了，而托勒密居然发现有必要使用一种更为可疑的设计——**对点**——为的是按简约和正确的方式来拯救行星运动的"表

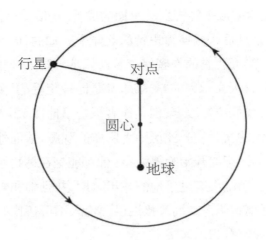

图8　对点是(偏心)地球的镜像,并且我们假定从对点上看来,行星以匀速运动。所以,实际上行星运动是非匀速的。

象"。

　　在一个行星模型中,对点是偏心地球的对称点,位于相反的位置。行星则被要求在它的圆上运动,使得从"对点"上**看来**行星在天空中以匀速运动。但是由于对点并**不**居于圆心,为实现匀速运动行星必须要改变其速率。托勒密是一个渴望知道所有时期行星位置的星相学家(他的《四书》是星相学的一部经典)。精确预报——不管所使用的方法是如何靠不住——较之认定所有在圆上的运动必定是匀速的这个哲学定则来说,是优先要考虑的事情。他和巴比伦人一样,认为预报的精确性而不是定则才是首先要考虑的。

　　开普勒的行星运动定律向我们揭示了为什么对点是如此有用的几何工具。

　　依照头两条定律,地球(或者其他行星)绕太阳按椭圆轨道

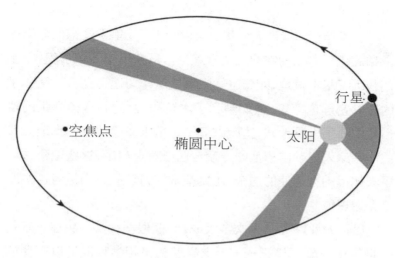

图 9 开普勒的头两条定律使得我们了解了为什么对点是一个有用的工具。它们意味着，一个在椭圆上绕日运转的行星在靠近太阳时运动较快，而在靠近椭圆的另一焦点时运动较慢。结果，从这"空"焦点上看起来，行星的运动是近乎匀速的。在本图中，轨道的椭率被极大地夸大了。

运行，太阳位于两个焦点之一，从太阳至地球的连线在相等时间里扫过相等的面积。所以，地球在其轨道上接近太阳时，运动就会加快，地球远离太阳（从而靠近椭圆的另一个"空"焦点）时，运动就会减慢。从空焦点看，地球穿越天空的速率将会是近乎匀速的：当地球接近太阳而远离空焦点时，地球要运动得快一些，不过由于地球与空焦点的距离较远，所以并不明显；当地球靠近空焦点时，地球要运动得慢一些，不过由于它与空焦点的距离较近，所以同样不太明显。换句话说，开普勒教导我们，近似地说，地球穿越天空的速率由空焦点看来的确是近乎匀速的。因此开普勒椭圆中的空焦点和托勒密圆中的对点是相对应的。

在中世纪晚期的大学里,学生学到了亚里士多德的哲学理论和简化了的托勒密的天文学理论。从亚里士多德那里,学生们获悉了基本真理,即天空绕中央地球作匀速运动。从简化了的托勒密理论那里,他们学到了本轮和偏心圆,这样产生的轨道的中心不再是地球,这就破坏了亚里士多德的基本真理。那些能够深入托勒密模型的专家会遇上对点理论,这些理论破坏了天体运动是匀速的这个(还要基本的)真理。哥白尼对这些理论尤为震惊。

尽管如此,利用《天文学大成》的模型——其参数在未来几个世纪中得到了修正——天文学家和星相学家能够以简便的方式和合理的精度计算行星的未来位置。异常的情况也存在。例如,在托勒密模型中,月球的视直径会有明显的变化,而实际上并非如此。该模型使用一种粗糙的特别设计,使天空中金星和水星始终接近太阳。但是,作为一本指导行星表制作的几何读本,《天文学大成》非常有用,而这才是其价值所在。

在《天文学大成》之后的著作《行星假设》中,托勒密提出了他的宇宙论。如同早先的希腊宇宙学家那样,他假设一颗行星在天空中相对于恒星运动的时间越长,亦即行星的运动和恒星的规则周日运动的差别越小,该行星离恒星就越近。如其如此,则有 30 年周期的土星是靠恒星最近和距离地球最远的,下面依次是木星(12 年)和火星(2 年)。月球(1 个月)距离地球最近。但是,在恒星之间结伴运转从而全都具有相同的 1 年周期的太阳、金星和水星,其位置又如何呢?因为在天空中太阳居于支配地位,并且有的行星与之结伴运行而其余行星则不随侍太阳,所以传统上认为太阳居于 7 颗行星的中间,直接地位于火星下方并且将随侍的行星和不与之结伴的行星分隔开来。金星和水星的位置长期有争议，托勒密将水星置于金星的下方，未必比掷一枚硬币有更多的依据。

　　依据各式各样的推理，从似是而非的到纯属猜测的，托勒密建立了行星的序列。托勒密现在假设，地球上方每个可能的高度不时地被唯一一颗特别的行星所占据。例如，月球的最大高度（托勒密有论据表明，这个值是 64 个地球半径）和最贴近的行星水星的最小高度相等。水星的几何模型确定了它的最大高度和最小高度之间的比率，这个比率再乘上 64 倍地球半径就得出了水星的最大高度。就是说，几何模型给出了每颗行星最小和最大高度之间的比率，而具有 64 倍地球半径的月球的最大高度则标定了整个系统。位于土星最大高度处的固定恒星在我们上方 19865 个地球半径或 7500 万英里处：托勒密的宇宙是令人印象深刻的大宇宙。

　　喜帕恰斯开始使用巴比伦提供的传统工具——算术的广泛用途以及长达几个世纪的观测定出的极其精确的参数——来追求希腊几何天文学的中心目标：依靠基于匀速圆周运动这一基本宇宙学原理而建立起来的一个几何模型再现每颗行星的完整轨道。托勒密完成了这项工作并达到了目的，虽然其中也做出了某些让步。《天文学大成》的模型在后来被屡次改进，但是只有在 14 个世纪之后，印刷术的发明才使得一个具有同等能力的数学天文学家认为其缺陷是根本性的，需要加以革新。

第三章
# 中世纪的天文学

　　622 年,先知穆罕默德迁出麦加到了麦地那,不久之后,新宗教伊斯兰教传遍了整个北非并传入了西班牙。伊斯兰教对天文学家的技能有特定的要求。每个月的起始是新月——不是当太阳、月球和地球排成一线时,而是在之后的两三天,当肉眼能看到蛾眉月的时候。可以使相邻的村庄都同意这个时候是新的一月的开始吗(即使天空云层密布)?祈祷的时刻是按照太阳穿越天空时的地平高度而定的,因而正确定出这些时刻的需求最后导致了"穆瓦奇特"(清真寺"授时者")官署的设立,给了天文学家一个稳定且受人尊敬的社会地位。当地的麦加方向即"奇布拉"的确定支配着清真寺和墓地以及其他诸多事务的形制,提出了一个具有挑战性的问题,这个问题正是"穆瓦奇特"和天文学家想要解决的。

　　在伊斯兰教传入很久以前,亚历山大在动荡年代就已不再是伟大的学习中心。《天文学大成》进入了君士坦丁堡,9 世纪时,来自巴格达的使者购买了一个抄本,使巴格达年轻而活跃的穆斯林文化认识到了以希腊语留传下来的知识宝库的重要价值。在巴格达,智慧宫的一个小组作了翻译,先从希腊文本译成叙利亚文,再从叙利亚文译成阿拉伯文。君士坦丁堡的其他抄本则被尘封而无人阅读，直到 12 世纪，皇帝将一个抄本作为礼物送给了西西里国王，并在那里被译成拉丁文。

占星术尽管在《古兰经》中受到批判，但仍然盛行于穆斯林世界的每一个社会阶层。那些并非仅仅是算命人的占星家将他们的预言建立在行星位置表之上。《天文学大成》模型的成功是无可争辩的，但这些模型所结合的参数在几个世纪后却被日益精确地确定出来——托勒密本人就曾说明过该怎样确定。最初，天文学家为此目的所用的天文仪器并不大，但是，随着观测者抱负的增大，观测仪器的尺寸也变大了，而且观测者指望赞助人来支付建造费用，并为他们提供住处。

但是，有时这会招致宗教权威的敌意，而一个赞助人的死亡——或者甚至是他勇气的丧失——也会引发天文观测的终止。在开罗，维齐尔①命令于 1120 年开始建造一座天文台，但是到了 1125 年，他的继任者却被哈里发下令杀死，他的罪名包括"与土星交往"，于是，天文台被拆毁。在伊斯坦布尔，土耳其苏丹穆拉德三世在 1577 年为天文学家塔奇丁建成了一座天文台——其时正值一颗亮彗星出现。塔奇丁无疑是为了自己的发达而将这一天象解释为苏丹和波斯人作战的吉兆，但是实际情况却正好相反。1580 年，宗教领袖使苏丹相信，窥探自然的秘密会招致不幸。苏丹于是下令将天文台"从远地点到近地点"彻底摧毁。

只有两座伊斯兰天文台存在的时间稍长一些。第一座在马拉盖，即今天的伊朗北部。这是波斯的蒙古统治者旭烈兀为杰出的波斯天文学家图西（1201—1274）在 1259 年开始建造的。

---

① 伊斯兰国家的高官大臣。

它的仪器包括一座半径为 14 英尺的墙象限仪（一种固定在正南北方向的墙上，用于测量地平高度的仪器）和一座环半径为 5 英尺的浑仪（用于其他的位置测量）。借助于这些仪器，一组天文学家在 1271 年完成了一部《积尺》①，即根据托勒密的《便捷表》传统而编制的天文表集，包括对使用方法的说明。但 1274 年，图西离开了马拉盖前往巴格达。虽然天文台的观测一直持续到下一个世纪，但是它的创造性时期却已经结束。

另一座主要的伊斯兰天文台受益于王子本人就是天文台的一位热心成员。在中亚的撒马尔罕，乌鲁伯格（1394—1449）在 1447 年继承王位之前就是一个行省的统治者，他于 1420 年开始建造一座 3 层的天文台，其主要仪器是按照"越大越好"的原则制造的一架半径不下于 130 英尺的六分仪。该仪器被装在户外两堵南北方向的大理石墙上，仪器的活动范围经过调节可以用来观测太阳、月球和五大行星的中天。撒马尔罕天文台的伟大成就是一套天文用表，其中包括一张有一千多颗星的恒星表。更早以前，巴格达天文学家苏菲（903—986）曾修正了托勒密的星表，他给出了改进过的星等和阿拉伯文的称谓，但是恒星本身和它们常常不正确的相对位置却没有得到修订。所以乌鲁伯格的星表堪称是中世纪唯一重要的星表。撒马尔罕天文台在乌鲁伯格于 1449 年被谋杀后很快就被废弃了。

天文台是为杰出人物建造的，但是每一位星相学家也需要作观测，这在研制出星盘以后成为可能。星盘是源于古代的一个精巧的便携式计算装置和观测仪器。典型的星盘由一个黄铜圆面构成，通过位于顶端边缘的一个圆环可以被悬挂起来。星

---

① 元代译名，指历表。

盘的一面可用来观测恒星或行星的地平高度,观测者将仪器悬挂起来并沿着一根瞄准杆观测天体,然后沿着圆周在刻度盘上读出该天体的地平高度。圆盘的另一面代表从天球南极投影到天球赤道面上的天球。

　　发自天球南极的每一根线与天球相交于一点,并且与赤道平面相交(于一个唯一的点),后者是前者的投影。因为黄铜盘

图 10　在牛津大学莫顿学院保存的一个 14 世纪的星盘。

面的尺度是有限的,又因为当时的星相学家对于南回归线以南的星空没有实际兴趣,故而投影的天空从天球北极(以圆盘中心来代表)一直延伸到南回归线为止,不再向南。

在观测者纬度处的等高度圈投影成的圆,和很多别的东西一起,被蚀刻在盘面上。到目前为止,一切都好,但是转动天空的恒星也需要显示。这可以通过一个黄铜薄片来完成,它指示出主要恒星的位置,并且尽可能多地挖去其余部分,从而使得下面的坐标圈露出来。这个黄铜薄片绕着下面一个圆盘的中心转动,就像恒星绕着北天极转动那样,它也能显示出太阳的黄道轨迹,观测者需要知道(并且标出)太阳在黄道轨迹上面的现时位置。

这样一来,在一次观测中——典型的观测包括在夜间一颗恒星的地平高度观测或在白天太阳的地平高度观测——观测者可以将该黄铜薄片转动到它正确的现时位置,亦即移动它直至恒星(或太阳)位于适当的坐标圈之上。至此,整个天球现在就到位了,而且许多问题可以获得解答——例如,哪些恒星现在正处于地平线上方以及每颗恒星的地平高度是多少。将太阳和圆盘周边上的刻度连成一线并在标尺上读出钟点,即可确定时间。在确定黄铜薄片的位置以后,无论观测的是恒星还是太阳,星盘总是像钟一样,能够日夜24小时告知时间。

我们能够从星盘容易地获取大量的其他信息。例如,确定一颗恒星升起的时间。天文学家可以转动黄铜薄片直至该恒星位于零地平高度圈之上,然后读出时间就行了。星盘是一个简单、精巧而又多用途的设计,它促进了天空的定量观测。

早在 9 世纪的上半叶, 巴格达智慧宫的花剌子米[al-Khwarizmi,此人名字的讹误拼法给了我们"算法"(algorithm)

这个词]就编制了一部《积尺》。它使用了桑斯克里特天文著作中所包含的参数以及计算过程。770 年左右，桑斯克里特的这本著作就被带到了巴格达。《积尺》于 12 世纪被译成拉丁文，从而变成了印度天文学方法运抵西方的载体。《积尺》使得对未来行星位置的预报成为可能，从事职业活动的天文学家或星相学家也应运而生，于是这类星表大量产生了，并且其中常常用到改进了的托勒密参数。

伊斯兰世界没有在基督教西方出现的大学的对应物，我们试图寻找一位有创意的挑战亚里士多德或托勒密地心宇宙基础的伊斯兰思想家，但却一无所获。在 10 世纪，经常出现怀疑托勒密的讨论。受攻击最多的是托勒密的对点，它违反了匀速圆周运动这个基本原理。但是本轮和偏心也遭到了批评，因为有关的运动虽然是匀速的，但却不再以地球为中心。这方面的一个纯粹主义者是安达卢西亚人拉什德（1126—1198），他在拉丁世界为人熟知的名字是 Averroes，在西方，亚里士多德被称为"哲学家"，而 Averroes 则为"评注者"。他承认托勒密模型"拯救了表象"——再现了观测到的行星运动，但是这并不意味着模型就是真实。他的同事安达卢西亚人比特鲁基（他的拉丁名字是 Alpetragius）尝试设计了替代模型，以满足亚里士多德学派的需要，但其结果当然令人很不满意。

在开罗，哈桑（965—约1040）试图修改托勒密模型，使得它们具有物理实在性的特征。在他的《论世界的构造》一书中，天空由同心的球壳壳层构成，在各层的最厚处，则有更小的球壳和球。在 13 世纪，他的著作被译成了拉丁文，并成为在 15 世纪影响乔治·普尔巴赫的著作之一。

甚至在最具实用观念的天文学家中间，对点也早已引起了疑虑。13 世纪,马拉盖的图西成功设计了一种含有两个小本轮的几何替代物;出于同样的原因,哥白尼在他生涯的某个阶段也采用了一种相似的设计。不过历史学家还没有发现他们之间有明显的联系。14 世纪中叶,大马士革倭马亚清真寺的"穆瓦奇特"伊本·舍德尝试设计了行星模型,剔除了所有引起异议的元素。他的月球运动模型避免了《天文学大成》中月亮视直径的巨大变化,他的太阳运动模型基于对太阳的新的观测,他的全部模型不仅摆脱了对点,而且也摆脱了偏心圆。但是他发现,由于我们能够很好地理解的理由,本轮是不可避免的。不过在舍德的时代,拉丁世界就已经发展起了它自己的天文学传统,从而不再依赖阿拉伯文的翻译。

这种独立是缓慢形成的,在罗马世界,没有一种主要的古天文著作是用拉丁文撰写的,希腊语仍然是学者的语言。随着罗马帝国的崩溃,希腊的知识在西方几乎全部消失,于是人们不再阅读古天文学的经典——即使是可以获得的。罗马哥特王国的高官博埃修斯(约 480—524/5)着手将柏拉图和亚里士多德的论著尽可能多地译成拉丁文,但是为时已经太晚。不过不管怎样,在因一桩不公正而反抗国王从而被处死以前,博埃修斯设法翻译了许多希腊语著作,其中有几种是逻辑学著作。他将这些著作和西塞罗这样的罗马作者的逻辑学著述放在一起,从而留赠给后世一套文集,这套文集成为长期研究的一个领域,中世纪的学生可以在书中进行"比较和对照"并且得出自己的结论。结果,逻辑学上的相容性成为了中世纪大学里的一大议题。关于本轮的真实性以及行星模型能否在根本上达到确定性的辩论,成为了人文学科青年学生们的兴趣所在。

在这段时期里,柏拉图只有一部(不完整的)著作被译成拉丁文：宇宙学神话《蒂迈欧篇》。卡西迪乌斯（4世纪或5世纪）翻译了该书的2/3，还写了一段冗长的评注。虽然地球为球形这个基本事实从来没有被忽视，但在中世纪早期用拉丁文写成的天文学著作读起来简直是糟透了。生活于5世纪早期的非洲人马克罗比乌斯为西塞罗的《西庇阿之梦》写了一个评注，在其中他阐述了一种宇宙学理论。在这个理论中，球形的地球位于布满恒星的天球的中心，该天球带动着行星每天自东向西地转动着，每颗行星也有着相反方向的自转，因为马克罗比乌斯的材料来源不一样，所以他对于行星序列的表述很模糊。迦太基人马丁纳斯·卡佩拉（约365—440）写了《哲学和墨丘利的婚礼》，这部著作是一个关于天堂婚礼的预言，在婚礼上，7个女傧相都献出了一门人文科学的纲要。这段描述对于解释以下问题很重要：为什么金星和水星总是出现在太阳的附近？其天文学解释是：它们绕太阳旋转，所以当太阳围绕地球旋转时，它们伴随着太阳。

像伊斯兰教那样，基督教也对天文学家提出了挑战，其中主要的挑战是计算复活节的日期。简而言之，复活节是春分第二个满日之后的那个星期天。这样，它在任一年的日期就同时依赖于太阳和月球的周期。作为巴比伦人传下来的月和年精确值的继承者，亚历山大的基督教徒可能提前若干年就已算出了复活节的合适日期，但是教会的权威人士则采取了更为实用的方法，尝试着找出一个由许多年构成的期限，它几乎和一个月的整数倍相等，并可设定未来年份中复活节的日期。这个日期一旦确定，这样一个序列就可以在未来的周期中年复一年地重复。

最后采用的周期是由巴比伦天文学家在公元前5世纪发

现的,但却归功于希腊人默冬了,这一周期的依据是 235 个朔望月等于 19 年（误差只有两个小时）。725 年,英格兰贾罗的"可敬者"比德(672/673—735)写出了一篇关键性的论文《论时间的划分》。在恺撒制定的儒略历中,每 4 年就有 1 个闰年（无一例外）,所以每 4 年,一个给定日的周日总是超前 5 天,从而在 7 × 4=28 年后,周日将回到原先的日期。比德将这与 19 年的默冬周期结合在一起,算出一个总的周期 19 × 28=532 年,既迎合了复活节与日和月的配合关系,又满足了复活节要在星期日的需要。

天文学以及占星术在拉丁世界的复兴是在第一个千年的末期开始的。当时星盘从信仰伊斯兰教的西班牙传入了西方。在那些日子里,占星术有一个合理的基础：植根于微观宇宙——单个生命体——和宏观宇宙之间的亚里士多德哲学类比。医科学生学会了怎样追踪行星,这样他们就知道什么时候有利于治疗病人的相应器官了。

1085 年,伟大的穆斯林中心托莱多陷入了基督教徒之手,伊斯兰的知识宝库,特别是希腊文变得可以理解了。翻译家移居到西班牙,最著名的是克雷莫纳的杰拉尔德（约 1114—1187）,他数不清的译作中包括有《天文学大成》和撒迦里的《托莱多天文表》(1100)。这些表经修改后被其他地方所采用,事实证明它们很好用,虽然构成其基础的行星模型暂时还是个谜。

如果说 12 世纪是翻译的时代,那么 13 世纪就是译作被吸收的时代。在涌现的大学里,拉丁语是法兰西式语言,所以没有语言上的障碍阻止学生和教师去他们想去的地方。未来的律师可能到博洛尼亚,医科学生可能到博杜瓦,而对大多数学科来说,巴黎是个理想的去处。

在那里,像在别的地方一样,人文学科学院通过人文学科的 7 门课程(其中包括天文学)提供文学和计算方面的基本教育。人文学科的学生大多为十多岁的男孩子,而印刷术还未发明,所以讲授的水平不可避免地只是初等的。少数学生最后会留在更高级的学院,从事神学、医学或法律研究。医学和法律享有传统声誉,奥古斯丁和教会其他神父的著作确保了神学是门有前途的学科。所以,在更高级的学院教师和陷入人文学科单调常规的教师之间存在着一种紧张的关系。

大多数新的译作属于人文学科,阅读这些译作为巴黎的文学硕士提供了一种提高地位的途径。同时,亚里士多德文集的传播对于基督教《启示录》并无贡献,甚至似乎还挑战了某些基本的基督教教义,在神学家之中引起了疑虑和不安。紧接着是巴黎的十年混乱,直至多明我教会的修士托马斯·阿奎那(1225—1274)完成了综合,成功地将亚里士多德吸收到基督教教义之中,以至于 17 世纪的人们发现很难将两者分开。

研究并不是那时候大学的任务,在天文学中最要紧的教学需要是一本青年学生可以用的初等教科书。13 世纪中叶,霍利伍德的约翰——他的拉丁名字是 Johannis de Sacrobosco——在这方面作了尝试,但是他的《天球》在解释太阳、月球和行星的运动时遇到了挑战,故而并不合适。尽管如此,在印刷术发明之后,这本著作还是为更有能力的天文学家提供了一个借口,去写出详尽的评注,并且就以这种形式成为了所有时代的畅销书之一。

在 13 世纪后期有个匿名的作者对《天球》连同他自己的《行星理论》的某些缺点作了校正。这给了各式各样行星的托勒密模型一个简单的(即使并不令人十分满意的)说明以及清晰的定义。同时,在卡斯提尔国王阿尔方索十世的宫廷,旧的《托莱多天文表》被《阿尔方索星表》所替代。现代计算机分析表明,这些在以后300年中被视为标准的星表是依据托勒密模型计算的,只是略有参数更新而已。

到14世纪,西方拉丁世界在充分掌握了过去的传统后,有了新的突破。对天文学来说,一个有意义的发展来自地球物理学。通过讨论抛体运动,亚里士多德曾经令人信服地论证了地球是静止的:垂直射出的箭会落回地面,正好是射手放箭的位置,这证明在箭飞行时,地球并没有运动。

但是,在讨论抛体运动时,亚里士多德并不能令人信服。他论证道,像一支箭那样的地球上的一个物体将会自然地向着地心运动,而其向上的(因而是非自然的)运动必定是由一个外部的力施加其上造成的——并且不只是施加其上。只要箭还在上升,这个力就一直存在。亚里士多德认为空气本身是维持箭向上运动的唯一动因,但是怀疑论者指出这是似是而非的,因为箭有可能会不顾大风而向上射去。

巴黎的大师让·布里丹(约1295—约1358)和尼古拉·奥莱斯姆(约1320—1382)同意亚里士多德关于必有外力作用的观点,但是他们反对将空气视作一种外力。他们表示,弓箭手必将一种"无形的动力"施加在箭上,他们称之为"冲力"。布里丹认为天球——它们虽然没有摩擦力,但是要永远地自转下去则需要有个持久的动力(天使的智力?)——只要在创世时被赋予促成运动的冲力,就会永恒地旋转下去。

奥莱斯姆明白冲力概念的重要含义,**如果**地球的确是自转的,那么站在地球表面上的弓箭手会和地球一起运动。结果,当他准备射出箭时,会不知不觉地向箭施加侧向冲力。在这个冲力的作用下,飞行中的箭会水平地移动,也会垂直地移动,保持与地球同步,从而正好回落到它射出的地方。他说,因此箭的飞行对于地球是否静止的争论起不了作用。其他援引的传统论证,包括出自基督教《圣经》的引文,也证明不了什么。奥莱斯姆的见解是:地球的确是静止的;但这也不过是个见解而已。

15世纪印刷术的发明有许多影响,最重要的是促进了数学学科的发展。所有抄写员都是人,在准备一个原稿的副本时都会犯偶然的错误。这些错误常常会传递给副本的副本。但若著作是文字作品,后来的抄写员注意到了文本的意义,他们就有可能识别和改正他们的同行新引入的许多差错。但是那些需要复制含有大量数学符号文本的抄写员难得运用这样的控制手段。结果,撰写数学或天文学论文的中世纪学生会面临巨大的挑战,因为他只有一个手稿的副本可用,而该副本不可避免地在传抄中会有讹误。

在引入印刷术之后,所有这些都改变了。现在作者或译者能够核对校样,从而确保排好字的文本忠实反映出了他的意图;然后印刷机能够印出许多完美的副本,用来分发到整个欧洲并且出售。与手抄本的费用比起来,印刷本的价格也更为适中。

在几十年内,希腊天文学家的成就就已经被掌握并且还被超越了。奥地利宫廷星相学家乔治·普尔巴赫(1423—1461)的《行星新理论》于1474年付印,它描述了支撑《阿尔方索星表》的托勒密模型和这些模型的真实物理表现,也许就是这些表现的不足激发了哥白尼去从事天文学的研究。

1460年,普尔巴赫和他的年轻合作者柯尼斯堡[拉丁文为Regiomontanus(雷纪奥蒙塔努斯)]的约翰尼斯·穆勒(1436—

1476）遇到了尊贵的君士坦丁堡的红衣主教约翰尼斯·贝萨瑞恩（约1395—1472）。贝萨瑞恩渴望看到《天文学大成》的内容更易理解，他说服了这两位天文学家着手完成这一任务，普尔巴赫在次年即告离世，但雷纪奥蒙塔努斯[①]完成了这一任务。他们的《天文学大成梗概》篇幅只有原著的一半，于1496年以印刷本面世。它至今仍是托勒密名著的最佳简介之一，至于《天文学大成》本身则于1515年以过时的拉丁文译本问世，1528年出版了新的译本，1538年出版了希腊原文版。到了1543年，一本胜过它的书出现了。

尼古拉斯·哥白尼（1473—1543）出生于波兰的托伦，就读于克拉科夫大学，在那里天文学教授从不隐瞒他们对于对点概念的不满。然后他去了意大利，并在那里学习教会法规和医学，同时也学习希腊语并发展他在天文学方面的兴趣。据说，1500年左右，他曾对大批听众作过天文学讲座。1503年，他回到了波兰从事弗龙堡大教堂的行政管理，他的舅舅是那里的主教。他的余生就待在这个主教管区。

中世纪晚期，亚里士多德卷帙浩繁的著作在拉丁世界是可以读到的，但柏拉图的境遇就不怎么好了：只有两篇不重要的对话添加进卡西迪乌斯很久以前曾作部分翻译的《蒂迈欧篇》里。这些状况在文艺复兴时期改变了，因为此时恢复了与希腊世界的密切接触，在君士坦丁堡1453年陷落之前不少希腊学者涌入西方。柏拉图的对话被发现并以其文学价值而备受赞赏，而且他对宇宙的数学方面的观点开始取代自然主义者亚里士多德的宇宙理论。天文学家开始寻找行星理论中的和谐和匀

---

① 即穆勒，他以其出生地的拉丁文名字闻名于世。

称。但是他们没有在托勒密模型那里找到，即使《阿尔方索星表》仍能以合理的精度满足需要。他们对对点理论尤为不满，用哥白尼的追随者乔治·约阿希姆·赖蒂库斯（1514—1574）的话来说，对点理论是"自然所憎恶之事"。

托勒密的《行星假设》连同他总的宇宙论现在也已失传。《天文学大成》提供的是单颗行星的模型，但是托勒密明显没能呈现出一幅宇宙的整体图像。正如哥白尼所写的那样，这意味着过去的天文学家

> 未能发现或推断出所有的要点——宇宙的结构和它各部分的真正对称。但是他们，正如有些人那样，从不同的地方取来了手、足、头和其他肢体，描绘得的确很好，但造型不取自同一身体，相互之间也不匹配——故而这样的部件组合起来，与其说是个人，倒不如说是个怪物。

正是基于这些美学上的考虑以及托勒密模型月球视直径的荒谬变化之类的许多特殊问题，改革的压力产生了——即使托勒密模型（连同改进了的参数）已足够合理。

改革的方向有迹可寻。亚里士多德的习惯是引述他打算予以反驳的那些人的话，这意味着每个学生都得了解那些曾主张过地球是运动的这一观点的古代作者——亚里士多德的反驳已经不再令人信服。普尔巴赫曾评述过，由于未知的某种原因，在每一个单颗行星的模型中如何会产生一个年周期。无论是什么驱动了哥白尼的思想，在他返回意大利没有多少年之后，一本由他撰写的题为《要释》的手稿开始流行。在其中他简述了他对现存行星模型的不满，对于对点理论更是有专门的批评。他

提出了一个以太阳为中心的替代方案,其中地球变成了一颗周期为一年的围绕太阳转动的行星,而月球则失去了行星的地位,成为地球的一颗卫星。

他指出这将如何为行星(现在的数目为 6 个)排列一个明显的序列(按周期和距离)。我们知道托勒密如何似是而非地假定,运动最慢的行星是天空中最高的;但这没有确定太阳、水星和金星间的序列,当它们相对于恒星运动时,它们结伴而行,所以看起来周期同为一年。一旦这个周期被认为是地面观测者的实际观测周期,就能找出水星和金星的真正周期,这两个周期非常不同,也和地球的周期不同,因而可以由此建立一个明确的周期序列。

哥白尼也能用地球与太阳距离的倍数来度量行星轨道的半径:例如当金星看起来距离太阳最远时(位于"最大距"),地球—金星—太阳构成一个直角,通过测量金星—地球—太阳间构成的角度,观测者能够画出这个三角形的形状从而得到其边长之比。周期序列和距离序列被证明是统一的。他后来说到了这一点:

> 因此,在这种安排中,我们发现世界有一种奇妙的相称性,并且在运动的和谐和轨道圆的数量之间有一种确实的联系。这是用任何其他方式都无法发现的。

在那个时代,人们都是在像托勒密那样寻找宇宙中的和谐,因此以上的观点是一个有力的观点。同时,哥白尼在《要释》中更为详细地率先发展了行星和月球的无对点模型。

岁月流逝,其间哥白尼在远离欧洲知识中心的地方发展着他的数学天文学。1539 年,当时还是维滕贝格大学的一名数学教授的赖蒂库斯拜访了他。赖蒂库斯发觉他自己已被哥白尼的成就迷住了。哥白尼发展了行星运动的几何模型,足可匹敌《天文学大成》中的模型,但是它们被融入了一套完整的日心宇宙观中。他获得了哥白尼的许可,于次年发表了该著作的《简报》。他还说服哥白尼允许他将著作的完成稿[以简化的拉丁文书名 *De revolutionibus*(《天体运行论》)为大众所知]带到纽伦堡付印,并委派了路德派教士安德烈亚斯·奥西安德(1498—1552)照看印刷事务。出于好意,教士插入了一篇未经授权和不加署名的序言,提出这里太阳的运动只是一种方便的计算手段。结果那些只看序言的读者就对作者的真实意图一无所知。

　　哥白尼的书大量涉及行星轨道的(无对点)几何模型,而且以其使人气馁的复杂性让这些模型与《天文学大成》的模型相匹敌。后来人们才证明了它们能够成为精确行星表的基础——证实日心体系能够通过实践检验的伊拉斯谟·莱因霍尔德(1511—1553)的《普鲁士星表》于 1551 年出版,该星表正是用哥白尼的模型计算的。在《天体运行论》的第一卷中,哥白尼概述了突出的结果,那些结果都是依据地球是一颗围绕太阳旋转的普通行星这一基本假设取得的。

　　我们已经看到,按周期排序的行星表与按距离排序的行星表是相同的。同样显著的是行星的神秘"流浪"(行星由此得名)变成了从另一颗行星观测一颗行星的明显而可期的结果:当火星在天空中位于冲日方向时,看起来火星在退行,仅仅是因为那时地球从"内侧"赶上了它,至于为什么水星和金星这两颗行星总是在靠近太阳时可见,而其余 3 颗则在子夜时可以观测到,也不再有任何神秘之处了:水星和金星的轨道在地球轨道的内侧,而其余行星的轨道则位于地球轨道的外侧。

确实,作为唯一具有一颗卫星的行星,地球是反常的。"固定的"恒星确实也没有显现出周年视运动。如果从处于周年轨道的地球上观测,它们本应该有的(哥白尼反驳说,恒星在遥远的地方,所以它们的"周年视差"很小,以至于观测者不能观测到)。但是这些是细节。日心宇宙是一个真实的宇宙:

> 所有行星的中心是太阳,从这个位置,它可以一瞬间照亮整个宇宙。对于这座最美丽的神殿,谁能将这盏明灯安放到另外或更好的地方? 事实上有人将太阳称为宇宙

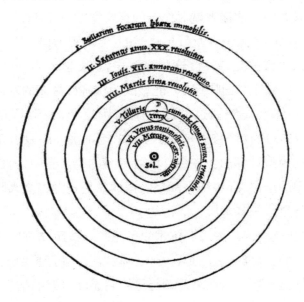

图 11 太阳系略图,取自哥白尼《天体运行论》第一卷。图中显示了各颗行星和它们的近似周期。注意第五颗是地球,是唯一带有一颗卫星(月球)的(哥白尼曾因这一异常而感到疑惑)。伽利略后来用望远镜发现木星也有卫星。

的明灯并无不当，另外有人称其为宇宙的心灵，更有人称其为宇宙的统治者。至尊的赫耳墨斯称太阳为"看得见的上帝"，索福克勒斯在《厄勒克特拉》中则称太阳为"万物的洞察者"。于是太阳好像端坐在御座之上，统治着围绕它旋转的行星家族。

希腊人曾试图对神秘行星"拯救表象"，《天体运行论》利用匀速转动的圆的组合这种几何模型将这样的尝试推向了极致。虽然《天体运行论》每个部分都和《天文学大成》一样复杂，但它是去掉了对点的《天文学大成》。几十年之后其革命性意义开始彰显。

中世纪的天文学

第四章
# 天文学的转变

　　哥白尼就他的目的以及着手实现其目的的方式而言,可能是传统的,但是他主张地球是运动的则提出了一整套的问题:是什么使地球运动?我们没有感觉到运动,是怎么回事?何以我们垂直向上射出的箭会回落到地面原先被射出的地方?如果我们观测恒星,当我们围绕太阳运行,每 6 个月从太阳的一边转到另一边时,为什么我们没有观测到恒星的视运动"周年视差"?我们又如何说明基督教《圣经》中那些似乎暗示了太阳在运动的段落?

　　有的人受《天体运行论》的未署名序言所误导,相信哥白尼自己并不主张地球是真正地在其轨道上运动;仅仅是为了更加成功地"拯救表象",他才使用了几何模型,其中地球被假想成运动的。另有些人——包括下一代几乎所有能干的数学天文学家——则专注于精心利用这些模型以拯救表象,从而忽视了哥白尼《天体运行论》的第一卷,在书中他清楚地说出了他的信条。又有些人再次寻求某种折中,其中有个人叫第谷·布拉赫(1546—1601),他在哥白尼创新的地方保守,而在哥白尼保守的地方创新。

　　第谷出身于一个丹麦贵族家庭,但是他并不按照封建社会里他的阶级成员所采取的方式生活，而是追求自己的学术爱好，其中天文学为他的最爱。1563 年发生了一次木星和土星的"合"。因为这两颗星是 5 颗行星中运动得最慢的，它们的"合"，亦即木星赶上土星，是罕有的事件，每 20 年才会发生一次，从星相学上讲是最不吉利的。十多岁的少年第谷在 1563 年"合"的时间前后对事件作了观测，并下结论说，13 世纪《阿尔方索星表》对它的预报超时一个月，即使是基于哥白尼模型的现代《普鲁士星表》也误差了几天，他认为这是不可接受的。不久之后，他就投身于观测天文学的改革。

像他的前辈那样，哥白尼对于用过去传下来的观测资料做研究甚为满足；只有当不可避免时，才会作新的观测而且使用的还是不尽人意的仪器。第谷的工作标志着观测天文学古代和现代的分界线。他将观测的精度视为建立好理论的基础。他梦想有一座天文台，可供他从事研究和开发精密仪器，并且有个熟练的助手团队负责测试仪器甚至编制一个观测资料库。利用他与高层的接触，他说服了丹麦国王弗雷德里克二世将汶岛授予了他，在那里，他于 1576 年至 1580 年间，建立了现代首家科学研究机构——乌兰尼堡（意为"天之城堡"）。

乌兰尼堡内一应俱全：4 个观测室，众多卧室，餐厅，图书馆，炼金室，印刷所。在岛上的别处，甚至还有个造纸厂，这样第谷在出版他的著作时就完全可以自给自足。4 年之后，第谷增加了他的设施，有了一个部分建于地下的卫星天文台"星堡"，其中的仪器和乌兰尼堡的不同，它们建在稳固的基座上并遮蔽了强风。但是，他不愿意承认乌兰尼堡有不尽人意之处，他说：再建一座天文台是为了阻止作平行观测的两队观测者之间互相串通。

图 12　乌兰尼堡第谷天文台中的大墙象限仪，墙是南北向的，观测者（在最右边，可勉强瞥见）在测量一个过子午线中天的天体高度。一名助手正大声叫出过中天的时刻，另一名助手正记录着观测。第谷和天文台各式人等的面貌被画在了墙上的图画中。

第谷留在汶岛直至 1597 年。那时弗雷德里克已被克里斯蒂安四世所接替,后者对第谷和他傲慢的举止表示不满,并给他的生活制造了日益增多的困难。于是第谷离职了。两年以后,他成为了布拉格皇帝鲁道夫二世的数学家。第谷已经失去对观测的热情;有的仪器留在汶岛,而其余则干脆存放了起来。但是他拥有对太阳、月球和行星的精确观测的大量资料,那是他的团队在汶岛获得的。这些资料被证明对开普勒的工作起了决定性的作用。作为助手,开普勒加入了第谷的团队,并且在第谷 1601 年逝世后,接替了他。

第谷是现代观测者中的第一人,在他含有 777 颗恒星的星表中,最亮恒星的位置已精确到 1 弧度左右。他自己可能最为他的宇宙论骄傲,但伽利略则并非是唯一一个将它视为是一种倒退的妥协的人。第谷欣赏日心行星模型的优点,但也意识到了对地球运动的异议,包括源自动力学、基督教经文和天文学的异议。尤其是即使使用他那高等的仪器,也观测不到周年视差,这暗示着哥白尼的辩解(恒星太遥远以至于不能观测到其周年视差)现在看来是非常没有道理的。根据他的计算,当恒星至少比土星远上 700 倍时,他才会出错,而且在行星和恒星之间如此浩瀚的、无目的的、空荡荡的空间实无意义可言。

因此,他寻找着一种宇宙论,它具有日心模型几何学的优点,但又坚持认为地球为物理上静止在宇宙中心的天体。解决办法在事后看似乎是显然的:使太阳(和月球)围绕居于中央的地球转动,并使 5 颗行星成为太阳的卫星。但是发现之路通常都是曲折的。1578 年,如同千年之前的马丁纳斯·卡佩拉那样,第谷在思考着使金星和水星成为太阳的卫星。到了 1584 年,他已将所有五大行星都变成了太阳的卫星,但这就意味着携带火星的球体会和携带太阳的球体相交。

也就在那时,第谷明白了他在 16 世纪 70 年代作的观测的意义。1572 年 11 月,一颗亮到足以白昼可见的恒星状天体出现在仙后座。有史以来,天空就被认为是没有变化的,但该天体却好像是颗亮的恒星。虽然他只有 26 岁而汶岛的工作还没有展开,但作为一个观测者,第谷已经取得了进步,可以确定该对象是天体而不是大气。其他人在这方面展开了争论,但是第谷将他们的观测作了一个判定性分析,最后解决了问题:天空的确能够改变。

有人可能认为彗星已经足够证明这一点了,但对亚里士多德学派的人来说,彗星的"来了又走了"充分证实了它们是天体(或者更准确地说,是大气)。正如亚里士多德本人曾说明过的,彗星是旋转的天空作用于围绕地球的大气和火而形成的,"所以每当圆周运动以任何方式搅动起原料时,在其最可燃之点就会突然燃起火焰"。

只要天空是不变的,就没有理由对亚里士多德声称彗星是大气表示异议。但是,对这颗新恒星的争议消失之后,第谷仍然抱有怀疑。只要大自然能提供他一颗明亮的彗星,他就可测量其高度并确定它究竟是大气抑或是天体。1577 年,当时乌兰尼堡还在建造中,大自然就赐予他一颗彗星;第谷证实了该彗星能在行星之间自由穿行。这表明,正如他稍后认识到的那样,被认为携带行星并以地球为中心的球体并不存在。

于是,对他的宇宙论的激烈反对意见消失了。在他那本 1588 年出版的论彗星的书中,他概略地提出了他的体系与太阳和月球运动的详细几何模型。

恒星直接位于土星的范围之外,与地球的距离约为地球半

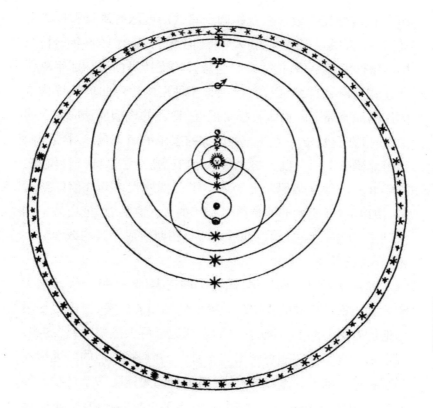

图 13 第谷宇宙体系略图。地球位于中心，太阳和月球围绕着它运行。地球本身被五大行星绕转，太阳带着它们绕地球旋转。恒星在最外面的行星土星之外。相对运动和在哥白尼体系中的情形一样。伽利略曾捍卫过哥白尼的理论，这一相对运动的情形给他造成了极大的困扰。

径的14000倍，所以第谷的宇宙甚至比托勒密的宇宙还要紧凑。

许多相似的折中方案在后来的几十年中浮现，而且其中许多还颇具吸引力。因为它们难以反驳，所以在支持哥白尼的论战中，它们激怒了伽利略·伽利莱（1564—1642）。作为 16 世纪

90 年代博杜瓦的数学教授，伽利略曾用地球的周日和周年运动试图说明令人迷惑的潮汐现象。但是，一直到 1609 年一桩戏剧性事件发生之后，他才全心全意开始支持哥白尼。那年夏天他在威尼斯，有传言说，在荷兰有一种用两片弯曲玻璃片构成的仪器能使遥远的物体看起来很近。弯曲的玻璃能造成变形的像，所以在交易市场上是一种传统的娱乐消遣工具。对传闻做了可靠确认之后，伽利略才着手为自己建造了这样一种仪器。同年 8 月，他向威尼斯当局展示了一架放大率为 8 倍的望远镜，使得他们"感到非常惊讶"。那年的 8 月之后，他把放大率增大到了 20 倍。再过数月，他就成为了图斯坎尼大公的数学家和哲学家，这并非巧合。

　　直到发明望远镜为止，像他们的前辈那样，每一代天文学家看到的都是相同的天空。如果他们知道的更多，那主要是因为他们有更多的书可读，更多的观测记录可供挖掘。所有这些，现在都改变了。在即将到来的岁月中，伽利略用他的望远镜看到了在他之前无人见过的奇观：自创世以来就隐匿于视线之外的恒星，围绕木星运转的 4 颗卫星，土星奇怪的附属物（半个世纪后才被认出是土星环），金星的似月位相，与地球上的山并非大不相同的月球上的山，甚至还有在设想为完美的太阳上的斑点。他能够证实亚里士多德的见解，认定银河是由不可计数的小的恒星构成。他发现肉眼所见的恒星看起来为盘状是一种光学幻觉。这样一来，如果信奉哥白尼学说的人为了避免在周年视差问题上的困难而被迫将恒星放逐到遥远的区域，那么他们就不必使它们在体积上巨大无比。伽利略能够如此谈论他的先辈："如果他们看到了我们之所见，他们也会做出我们所做出的判断。"从他的时代起，每一代天文学家都将比他们的前辈拥有更巨大的优势,因为他们拥有的设备允许他们去接近迄今未见的、未知的,因而未曾研究过的对象。

伽利略生活的那个时代里,科学期刊还未能为新著述的迅速传播提供论坛,以书的形式从容出版也还未能成为常态。但是他的望远镜发现不能等待,故而他在几个月以内在薄薄一本《星际使者》(1610)中宣告了他的最早发现。1613年《太阳黑子通讯》继而出版。他支持哥白尼的几个新发现,尤其是金星位相的似月序列,就是在这本书中宣布的。

在托勒密体系中,金星在行星序列里位于太阳的下方。此外,为了"确保"金星在天空中从不远离太阳,托勒密模型要求金星的本轮中心在从地球到太阳的直线上。结果在这个模型里,金星总是位于地球与太阳之间的某个地方。

如果金星被证明是个被太阳照亮的暗天体,那会怎样呢?如果托勒密模型是正确的,那么被照亮的一半将永远部分地背向地球,所以在我们看来,这颗行星永远不会像满月那样呈光轮状。相比之下,哥白尼认为,金星绕着太阳在地球轨道以内的一条轨道上运行。当它靠近地球时,会呈新月状,因为它被照亮的一半背向地球,情形如同托勒密模型中那样。但在远离太阳一边时,金星看起来会像满月一样。

这正是伽利略所目睹的。托勒密模型是错误的,它被一个决定性的观测所驳倒。因此哥白尼是对的——或者说,伽利略让我们相信他是对的。但是金星的位相告诉我们的仅仅是地球、太阳和金星的**相对**位置,并没有告诉我们它们三者之中哪一个是静止的。这种相对运动在哥白尼体系和第谷体系中是一样的。所以第谷体系并没有因伽利略的发现而受到影响。

对于伽利略而言,这是最不受欢迎的状况,因为托勒密长久以来就已经被那些相信地球是静止的、支持第谷或"半第谷"体系的人抛弃了。伽利略的望远镜发现使他成为了哥白尼学说的宣传员，但是证明托勒密是错的要比证明哥白尼是对的容易。于是他依然在托勒密和哥白尼之间进行非此即彼的验证。直至 1632 年，他给自己支持哥白尼学说的宣言取名为《关于托勒密和哥白尼两大世界体系的对话》。

为了消除物理学上的异议——当我们地球人在空间中飞驰时，怎么会始终相信我们事实上是在坚实的大地上？——伽利略创造了运动的一个新概念。运动——各种变化，位置的变化只是其中的一种——是亚里士多德哲学的基础，因为一个自然的物体是通过它的表现（它如何运动）来表达它的本性的。按照亚里士多德的观点，运动要求解释，而静止则不需要。

相反，伽利略提出了一种看待世界的新方式，其中运动的改变——加速度——需要解释，而定常运动（静止在此只是一种特殊情形）则是一种状态，不需要解释。他设想在球状地球完全光滑的表面上滚动着一个球，看不出有什么理由这个球应当回到静止：它会无限期地处于一种匀速运动的状态，绕着地球中心转动。相似地，地球本身绕太阳系中心运动，处于匀速运动状态，因此地球上的人感觉不到他们在运动。

伽利略既有交友的天赋，也有树敌的可能。他认为地球是运动的，这一观点长久以来被视为与基督教《圣经》的某些说法相矛盾。1613 年，伽利略给一位朋友写了一封半公开的信，这封信成为《致大公夫人克里斯蒂娜的信》（写于 1615 年，直至 1636 年才出版）的基础。现在它被认作是对传统天主教立场的经典陈述——《圣经》告诉人们如何去天堂，而不是告诉人们天堂如何运行——但是在反宗教改革时期，一个世俗人士很难发布对于《圣经》的解释。1614 年，一个传教士向他发动了攻击，把《使徒行传》中的一段经文转变成了也许是教会史上最好的双关语："加利利人呐，为什么你们站着呆呆地盯着天呢?"伽利略不顾朋友的告诫，坚持他的立场，争论升级并且梵蒂冈也卷了进来。最后在 1616 年 2 月，像圣徒一样的红衣主教罗伯特·贝拉明会见了他。贝拉明准备修改自己在地球稳定性方面的传统立场，但是只有在使人非信不可的证据出现的时候。他私下里让伽利略不可以再相信哥白尼体系是真实的或为其辩护。

时光流逝，1623 年曾经是伽利略的朋友和支持者的一个人被选举为新的教皇，此事鼓励他重新开始了支持哥白尼的运动，并且最终使得他的《对话》于 1632 年出版。究竟是什么激怒了罗马教会，这个问题仍处于争论之中，但结果是伽利略被判软禁在家。虽然对伽利略的软禁实际上并无残暴之处，但对他的定罪对天主教国家的天文学来说是一种挫折，许多耶稣会的天文学家不得不支持第谷体系或类似的折中方案。

在伽利略的性格中有着某种懒散，特别是他不太愿意去从事艰深的数学研究，这使他在宣扬哥白尼学说的运动中并未取得太好的效果，因为对于一个同时代人所做出的对哥白尼事业的贡献，他一直视而不见。这个人把行星视为受中心的巨大太阳产生的力所驱动的天体，从而将天文学从应用几何学转换成物理学的一个分支，将运动学转换成动力学。约翰内斯·开普勒（1571—1630）生于斯图加特附近的魏尔市，就读于蒂宾根大学，在那里，他那公正的天文学教师从正反两面讲述了包括哥白尼学说在内的各种各样的宇宙论。然后开普勒开始了神学研究。可是在 1593 年，当局任命他在格拉茨教数学，他勉强地照做了。

在格拉茨定居下来后，开普勒开始为宇宙的构造苦思冥想。他认为宇宙是由上帝创造的，上帝是个伟大的几何学家，他相信哥白尼已经发现了宇宙的基本布局，但却未能发现促使上帝从种种可能中选定这一宇宙的理由。特别是为什么上帝创造了正好 6 颗行星以及为什么他确定的行星之间的 5 个空间是现在这样的大小？最后，开普勒想到空间的数目等于正多面体（角锥体、立方体等）的数目，正多面体的形状必定对任何一个

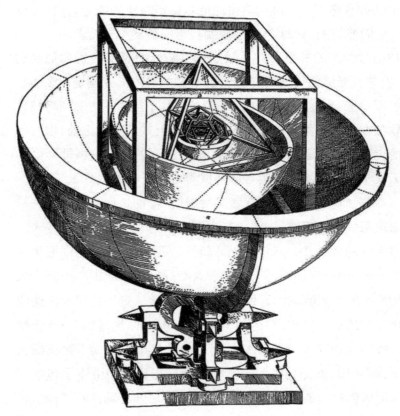

图 14 开普勒《宇宙的神秘》(1596)中论及的上帝在其宇宙中体现出的几何关系。

几何学家，无论是人还是神，有种极大的吸引力。因而开普勒开始调查研究由 6 个同心球组成的套件的几何构造，其中每一对球被 5 个正多面体之一以这样一种方式隔开：每对球的内球与正多面体的平面相切，外球则通过其顶点。最后，他找出了一个特别的序列，其中球的半径和哥白尼算出的结果相当一致。

这其中没有论及行星的速率问题，开普勒根据中心的——和巨大的——太阳的物理影响来探讨这个问题，这是划时代的一步。毕竟，哥白尼已经指出，一颗行星离太阳越远，它在轨道上就运动得越慢。也许这是因为是太阳导致了行星的运动，并且太阳作用力的有效性随距离增加而减少。

二十多年后，开普勒才建立了行星速率的实际模型，但是他发表于《宇宙的神秘》（1596）中的早先猜测已经向天文界宣告，一个新的天才人物登上了舞台。开普勒送了一本给伽利略，极力主张他出来支持哥白尼，但是他收到的只是一个礼貌的回应。尽管开普勒的书是第一本系统讲述日心学说的书，但第谷却邀请开普勒到汶岛来参加他的工作。开普勒决定不去那个遥远的岛屿，但是到了 1600 年初，当第谷重新定居在布拉格后，开普勒决定去探访他，并在那里作了 3 个月的火星轨道研究。水星有很多时间隐匿在太阳的强光中难以跟踪，除此之外，火星的轨道偏离圆最大，因此最难用圆形去"拯救"它。3 个月后，开普勒回到了格拉茨，但是他很快又回到第谷身边，第谷已经命不久矣。一年之内，开普勒就成为了第谷的继任者。

开普勒与战神火星的"战事"持续了好几年。他说，他的战役是基于哥白尼的日心理论、第谷·布拉赫的观测以及威廉·吉尔伯特（1544—1603）的磁哲学而展开的。吉尔伯特的《论磁》（1600）论证了地球本身是一个巨大的磁铁。

开普勒是那样另类，虽然是一个诚实的科学家，但在他的出版著作中并不理清他研究的过程，使得结论看起来直接又清楚。开普勒要求他的读者在计算的迷宫里追随他，而且他也没有合适的主题可作为引言，更不用说很简短的序言了。但是精华之点是明显的：开普勒放弃了传统天文学研究行星**怎样**运动的几何学模型，而是转向物理学，研究是什么力造成了行星的运动。

一旦第谷证实了通常被认为是携带行星运转的天球是不存在的之后，从运动学到动力学的转移几乎就不可避免了。人们对这些球为什么会继续转动已经少有兴趣——大多数人认为是天使的智力推动着它们。布里丹假设在创世时每一个球都得到了一个原动力，而哥白尼认为所有自然的球体自然地转动。但是在去掉了携带行星的天球以后，留下来的行星就是轨道上孤立的天体。那个驱动它们以近乎抛体方式运行的会是什么呢？一旦这个问题受到关注，日心假设的成功就有了保障，相对小的地球绕着拥有巨大质量的太阳运动（而不是反过来）就有了动力学意义。

1609 年，开普勒发表了他对火星问题的解决方案，该著作取了一个挑战性的题目，宣告了天文学的重新定位：《新天文学：基于原因或天体物理学，关于火星运动的有注释的论述》。之前，开普勒已经逐渐明白他必须采用真实的、有形的太阳作为他的太阳系的中心，而不是取某个几何上便利的点。同样，他必须对行星在经度和纬度上的运动作一个综合的说明。设计两个几何模型，每一个对应于某个坐标，而两者彼此不相容，这种办法已经不再能被接受了。

当太阳、地球和火星构成特别的位形时，第谷的资料很多，足以提供丰富的信息。例如当天空中发生火星冲日时,地面观测与从太阳系中心太阳上作的观测相似,第谷有许多这样的记录。这些记录是精确的:当开普勒用一个圆轨道模型"拯救表象"时,精度可达到 8 弧分(这个精度好到足以与此前任何一个观测者的观测相匹配)。开普勒知道,因为第谷的精度远比 8 弧分要高,所以这个模型还不能令人满意,因此他必须将它舍弃。但是第谷的观测(可说是)不是**太**精确:实际火星的轨道受到其他行星引力的扰动,因此这种具有理想精度的假设观测会妨碍开普勒提出以他名字命名的定律。

吉尔伯特曾证明地球是一个巨大的磁铁,也许太阳是更大的磁铁,因为行星全部绕太阳按同一方向公转,并且离太阳越远的行星运转得越慢。这使得开普勒将太阳看作一个转动的天体,它向空间发送磁力,推动着行星绕行,这种影响自然是对最近的行星最为有效。开普勒相信,如果没有这种连续的影响,行星就会在其轨道上停滞不前——也就是我们常说的惰性。

但是,因为行星的轨道并不是简单的圆,所以还有更多的问题。行星改变着它们的离日距离。为了解释这一点,开普勒在太阳中引入了第二个力,它在行星的部分轨道上吸引行星而在其余轨道上则推开行星。

在对第谷观测资料的分析中,这些物理直觉引导着开普勒。于是,开普勒发现了他的"第二定律",该定律告诉我们行星在其轨道上的速率,这比说明行星轨道是什么样的"第一定律"的提出还要早。根据第一定律,行星按椭圆轨道运动,太阳位于椭圆的一个焦点。公元前 2 世纪起,几何学家就熟知椭圆。不用充斥着本轮、偏心轮和对点的托勒密模型,而是通过一条异常熟悉的曲线就确定了行星的轨道,这堪称是一个妙举。但是在得出定律之际，开普勒的物理直觉在某个阶段变成了某种障碍——因为椭圆的短轴是对称的，而开普勒第一定律却告诉我们轨道的该轴是极不对称的，太阳位于一个焦点，而另一个焦点却是"空的"。

我们知道第二定律有个变形形式，牛顿后来发现这个变形形式是他的万有引力定律的一种**推论**：从太阳到行星的一条直线在相同时间里扫过相等的面积。数学家几乎无法处理这样一个古怪的表达式，他们宁愿选择其他形式，那些在数学上容易处理而且在观测上又几乎和面积定律难以区分的形式：行星运动的速率反比于太阳和行星间的距离；或者从空焦点看来，行星以明显的匀速率运动。在我们早先对圆中的托勒密对点的讨论中，我们知道为什么它在椭圆中的对应物——从空焦点上看到的匀速——与面积定律如此近似。

《新天文学》陈述了火星——由此可推知其他行星——怎样运动以及是什么推动它如此运动。但是，开普勒在《宇宙的神秘》一书中最先讨论过的行星系统综合样式到底有什么意义？这是《世界的和谐》（1619）一书中众多的论题之一，其他的论题还有行星在其轨道上产生的音乐。哥白尼高兴地发现，一颗行星离太阳越远，它完成一个回路的时间就越长。开普勒现在能够宣布一个公式来确保这一点：行星周期的平方与轨道半径的立方成固定的比率。

开普勒这时正努力使他的著作为广大读者所接受，将其设置为便利读者的问答形式。他的《哥白尼天文学纲要》曾在1618和1621年间陆陆续续刊行过，书名赞颂了他最大的灵感之源。但是哥白尼会困惑地发现，他的几何天文学在这里已转变成物理学的一个分支。

行星理论总是服从于最终的实际测试：它们能够用来生成精确的星表吗？第谷对天文学的兴趣是由采用哥白尼模型的《普鲁士星表》的缺陷所引起的，当第谷第一次将年轻的开普勒向鲁道夫皇帝引见时，皇帝委任开普勒与第谷一道制定出天文学家最终可以信赖的星表。1627 年，当鲁道夫和第谷已经去世很久之后，开普勒发表了《鲁道夫星表》。当法兰西天文学家皮埃尔·伽桑狄成为历史上第一个水星凌日的观测者时，开普勒也已死去。《鲁道夫星表》的预报要比《普鲁士星表》的预报精确 30 倍：开普勒的椭圆天文学通过了检验。然而，他的物理直觉（没有太阳不间断的推动，行星将会瞬间停滞）则是全然难以置信的。第二定律（确定行星在轨道上变动速率的关键定律）的表示法则会把人弄糊涂。开普勒用单纯的椭圆代替了众多的圆，并且他还鼓励天文学家将他们的学科视为"天体物理学"。但是，行星系统真正的动力是什么却仍是一如既往地神秘。

第五章

# 牛顿时代的天文学

　　中世纪后期的观点曾被亚里士多德所统治，文艺复兴时期的观点则由柏拉图所支配。但在随后的时代，"机械论哲学"或称"微粒论哲学"则变得越来越有吸引力。它源自希腊的原子论者，后者将我们对周遭物体不同的感知解释为我们的感觉对不变粒子运动的理解方式，这对一个时代是有吸引力的。这一代人发现用速度和形状等概念（至少大体上是数学概念）进行诠释，即可达到使人耳目一新的明晰。机械装置变得日趋精巧——斯特拉斯堡大教堂里的大钟就是明证。但是在这些机械中，复杂的效果是由简单的（而且明白易懂的）手段产生的：处于运动中的物体（齿轮、钟锤等等）。

　　上帝现在是伟大的钟表制造师，他创造的宇宙结构上非常复杂，但却像运动中的物体那样明白简单。伽利略被这种古代观念的复兴深深吸引，但是他更年轻的同代人勒内·笛卡尔（1596—1650）将这种机械论哲学推向了极致。在拉弗莱什的学校里，他的耶稣会教师在伽利略宣布望远镜发现后的几个月之内就向他介绍了这一发现。更重要的是，他们向他逐渐灌输了对于在几何定理中发现的确定性的极度赞美。对笛卡尔而言，在确定的真实和极可能的真实之间，存在着一条巨大的鸿沟。

他判定,为了创建两者之间的联结,人们必须仿效几何学家的推理。此外,几何学家对空间会有正确的理解:欧几里得无限、均匀、无差异的空间不是理想的抽象(人们曾经这样认为),而是真实世界的空间。

作为一位哲学家,笛卡尔是无情的。伽利略提及亚里士多德的时候会批判其观点。不同于伽利略,笛卡尔对亚里士多德不屑一顾而径自创建他自己的哲学。在此之前,所有的讨论,无论是赞成抑或是其他,都会回溯到亚里士多德。但此后不久,所有的讨论就会提及笛卡尔了。

在笛卡尔的宇宙里,不再有任何与寻常位置不一样的优越位置,诸如地球的中心或太阳系的中心(因为运动是围绕它们进行的)。在分析物质本身的基本概念时,笛卡尔摒弃了颜色或味觉之类的特性,因为它们只属于某些物质而不是所有物质,最终结论是物质和空间在许多方面是统一的。于是,没有物质的空间——真空——是不可能的:世界是充实的。此外,因为空间是均匀的,所以物质也是如此。这意味着我们在这个物质和那个物质之间所觉察到的差异完全是由(均匀的)物质在两个空间中的运动方式造成的:当我们着手理解宇宙时,运动就是一切。

因为上帝处在永恒的存在中,所以他使这个特别的物质在这个瞬间所具有的运动守恒,以给定的速度按一个特有的方向前进:直线惯性定律。正如他使宇宙的总体空间守恒一样,他使宇宙中运动的总量守恒。这使我们得以建立起支配运动从一个物体转移到另一个物体的定律。

因为宇宙中装填着物质,所以直线惯性与其说是一种实在,不如说是一种倾向。实际上,只有在其前面的(以及后面的)

物质也运动时,物质才能运动。结果是,物质通常作漩涡或旋转运动,这些漩涡像一架离心机,有的物质被驱向外部,有的则被推向中央。我们将后者视为自发光的,并且我们看到了像太阳和恒星那样的亮物质的巨大集合物。所以太阳只不过是离我们最近的恒星。在我们的无限宇宙中到处散布着相似的恒星。

太阳位于一个巨大的太阳漩涡的中央,这个漩涡携带着周遭的行星运转;这些行星有切向飞离之势,但却被漩涡中其他物质约束在封闭的轨道上。在行星中间有我们的地球,其本身就是携有月球的一个较小漩涡的中心。所以月球被太阳漩涡和地球漩涡两者运载着,因此,牛顿发现很难计算月球的运动。可是,虽然笛卡尔是一个数学家而且在一封信中他写道"我的物理学只是几何学",但在他表明观念的《哲学原理》(1644)一书中却只有文字而没有数学方程。文字是含糊而且圆通的,《原理》的适应性如此之强,以至于它几乎能够解释一切而又什么都不能预报。这本书能够被无数的人所理解,它所阐明的世界图景对巴黎沙龙里的热心人士具有强大的吸引力。

在这个世纪的后几十年里,亚里士多德在剑桥和牛津仍然居于正式的支配地位,虽然在个别学院中,有进取心的指导教师能够向他们的学生引入笛卡尔哲学的新颖内容。但是伦敦是建于 1660 年的皇家学会的所在地,院士们是曾经大大影响了开普勒思想的威廉·吉尔伯特"磁哲学"的继承人。约翰·威尔金斯(1614—1672)是皇家学会的领导人之一,他在 1640 年刊行了《月球世界的发现》第二版。在该书中,他论证了去月球上旅行在理论上是可能的,因为地球的磁影响随高度增加而减少:"这是可能的,磁力的大小反比于它距离磁源地球的距离。"

17 世纪 60 年代早期,皇家学会的实验室监理罗伯特·胡克

（1635—1703）甚至作了测试，看看地球的拉力在大教堂的顶端是否比在地面上小，结果当然是不确定的。胡克坚定地推广他关于磁学的想法，用于说明太阳系中我们周遭所见的现象。到了1674年，他能够用三个出色的"假设"来表达他已经达到的阶段：

> 第一个假设是，所有的天体，不管是怎样的，都有一种指向其中心的引力或重力，依靠这种力，它们不仅吸引自己的各个组成部分，阻止这些部分飞散（正如我们所观测到的地球那样），而且吸引位于引力作用球之内的其他天体。

胡克相信太阳系的所有天体以一种恒同于重力、维持地球各个组成部分在一起的力吸引着其他天体——更准确地说，是"它们作用球之内"的那些天体。关于直线惯性，他论述得异常清楚：

> 第二个假设是，已经在做直线和简单运动的天体，不管它们是怎样的天体，将会继续沿着直线向前运动，直到受到某种别的有效的作用力，才会偏斜或弯曲成用圆、椭圆或别的更复杂的曲线所描述的运动。
> 第三个假设是，引力的作用是如此强大，不管作用于其上的物质离得多近，引力仍然指向它们自身的中心。

但是，引力究竟是与距离本身成反比（$f \propto 1/r$），还是与距离的平方成反比（$f \propto 1/r^2$），抑或是别的什么形式？胡克不能够确定,他认为答案相对来说并不重要,仅仅是留给数学家的问题之一。

因为天体的亮度随距离的平方而衰减,故而反平方定律是明显的候选者。但是存在一个更加令人信服的理由。圆周运动(例如吊索上一块旋转着的石头)的动力学分析加上开普勒行星运动的第三定律使人联想起,如果行星全部以严格的圆周轨道和均匀的速率围绕太阳运动,这些轨道的全部样式都可以解释为太阳引力随距离平方减少的结果。但是行星的真实轨道是椭圆。能够证明椭圆轨道类似地也是由引力的反平方定律所引起的吗?

到了1684年,伦敦的意见变得坚定了:答案是反平方定律。但是无人能用数学证明这一点。才华横溢而又守口如瓶的剑桥数学家艾萨克·牛顿(1642—1727)能做到吗?埃德蒙·哈雷(约1656—1742)鼓足勇气挑战了牛顿。他问道,一颗行星被太阳按反平方定律吸引,其轨道将呈现什么形状?牛顿毫不犹豫地给了哈雷他所希望的回答:一个椭圆。

牛顿于1661年进入剑桥,在1665年成功分析了严格圆周运动的动力学。但是在试图了解行星的轨道时,他面临了严峻的问题:在笛卡尔的宇宙(牛顿那时候将它视为一个实在的真实)中,月球被太阳的漩涡和地球的漩涡两者所携带,因而使数学分析异常艰难。至于行星,有人则提出开普勒第二定律的几种变形,它们在观测上差不多是相同的,但概念上却有天壤之别。牛顿自己试图用该定律的对点形式进行研究,但是,最终他出版了一本书,将这个定律表达成我们今日所知的面积公式。

1679年,当收到一封来自胡克的信时,牛顿仍然致力于弄清漩涡的意义并且仍被轨道运动的动力学分析弄得一头雾水。

胡克现在是皇家学会的秘书,他渴望剑桥的数学家参与学会的活动。他邀请牛顿考虑"行星切向运动[惯性运动]和向心运动"的后果。胡克并不将轨道运动视为离心力和向心力之间斗争的结果,而是将其视为吸引力对运动的影响,否则会继续按直线运动。

胡克也告诉牛顿,他试图(见第六章)测量周年视差以证明地球的运动。在答复中,针对地球必定静止的传统"证据"(垂直向上射出的箭回落到发射的原地——或者等价地,从塔上丢下的一块石头回落到塔基的地面上),牛顿提出了新的方法。牛顿指出,因为塔顶比塔基距离地心更远,并且因为地球事实上是在自转着,所以仍然处于塔顶的石头比塔基地面的水平运动要快。于是他论证说,因为石头下坠时仍维持其水平速度,故而事实上它撞击的那块地面应该在塔的前方。他继续讨论这块石头会怎样运动:在假想情况下,石头可能不受阻碍地穿过地球。如此,牛顿将一个自由落体问题转化成了一个轨道运动问题。

胡克得意地指出了牛顿分析中的一个错误,尽管穿越地球的一条假想路径很对他的味口,但是他确实向牛顿表示过对反平方定律的支持:"我的推测是,引力永远与离中心的距离的平方成反比。"

对于任何一丝批评,牛顿的反应是不予理会,同时他悄悄地致力于建立物质的真实。虽然他认为胡克所提的方案脱离了充满物质的真实(笛卡尔的)宇宙,他还是进行了数学分析并且作出了非凡的发现:胡克的行星沿着以太阳为一个焦点的椭圆运动,从太阳到行星的连线在相等的时间里扫过相等的面积,恰似开普勒对于真实行星的描述。能否说胡克的世界是真实的——一个几乎为空的世界里,孤立的天体不知为什么能依靠引力穿越其中的空间而相互影响——而笛卡尔的充实反倒是虚幻的？

我们对于牛顿思想从 1679 年至 1684 年（哈雷的造访）的发展知之甚少，只知道牛顿与皇家天文学家约翰·弗拉姆斯蒂德（1646—1719）之间就 1680 年 11 月超近太阳的一颗彗星是否就是次月离开太阳的那颗彗星（确实是同一颗）有费解的信件往来。如其如此，其间发生了什么？又是为什么会发生？牛顿一方面认为它可能是一颗"从太阳的范围内流了出去"的彗星——它已经绕过了太阳的背面——并且他或许已经想到这颗彗星的路径是太阳引力的结果，但是我们对此不能肯定。当然，恭敬适度并且举止得体的哈雷的造访鼓励了牛顿，牛顿答应哈雷要证明椭圆轨道是由太阳的反平方引力所引起的。草稿越写越长，最终形成了《自然哲学的数学原理》（1687），该书以缩写的拉丁文标题 *Principia*（《原理》）更为人所熟知。同时代人因书名中对笛卡尔冗长而怪诞的《哲学原理》的挑战和非难而接受了它。

最初的草稿只有 9 页，它分析了在空的空间中以惯性运动的一个天体在一个"中心"的拉力影响下的轨道。这样一个天体会服从开普勒面积定律。如果拉力符合反平方定律，则轨道会是圆锥曲线——椭圆、抛物线或双曲线。如果天体按椭圆轨道运动，拉力指向其焦点，则轨道会服从开普勒第三定律；反之亦然。所有三条从观测中导出的开普勒定律（第二条则以"面积"形式表示），在一种高度可疑的动力之下，现在都已经被说明了是直线运动在反平方力作用下的结果。

牛顿还没有将引力视为大大小小的天体之间一种相互的作用力，如同更早些时候胡克所做的那样；这是难以理解的，因为他的草稿上说过开普勒第三定律可适用于木星的伽利略卫星以及土星的 5 颗卫星（克里斯蒂安·惠更斯在 1655 年发现了土卫六，而后卡西尼又发现了 4 颗卫星）。所有这些卫星被它们的母体行星所吸引，人们就觉得奇怪了：如果土星会拉土卫六，为什么土星却不会拉太阳呢？可能牛顿也有同样的想法，因为在下一个手稿中，引力就成为万有引力了。

　　牛顿认为，充满物质的笛卡尔的宇宙（其中的天体不停地相互撞击）现在已让位于几乎为空的宇宙（其中天体做直线惯性运动并受所有其余天体引力的影响——引力能穿越空的空间而抵达）。牛顿理所当然地被由之造成的数学挑战的复杂性吓呆了，尤其是在研究地球和太阳的共同拉力下的月球运动的时候。对欧洲大陆的数学家来说，他们因牛顿求助于一种神秘的"引力"这一后退之举而大受震动。对于这种引力，牛顿没有给出机理，它对整个世界而言像是重新引入了可疑的"同情心"以及机械论哲学刚刚才予以剔除的其他"隐秘的特性"。

　　在经过两千年的观测和分析之后，即使反平方力的性质依然神秘，行星轨道最终却被弄懂了。但是彗星的情况又如何呢？牛顿现已确定，1680 年末所见的两颗彗星是同一颗，而且它的确是从太阳周遭的范围内"流出"的。他断定，彗星是同一普适样式的一部分，而且在《原理》中他指出彗星的轨道是圆锥曲面（虽然不是必须为椭圆），而且它们也符合开普勒面积定律。这展示了以扁长椭圆轨道运动的彗星会定期地回到太阳系的可能性。

　　胡克久已怀疑地球对下坠石头的拉力与它对天体的拉力是一样的，现在牛顿也有同样的想法。但是将地球对一块石头的拉力和它对月球的拉力相比时，他面临着一种数学上的挑战。他将不得不把构成地球的所有物质对石头的拉力合并在一起,这些拉力作用的距离范围从几英尺到几千英里。将一种力通过薄薄的大气和通过若干英里的岩石和泥土后如何具有同等有效性的问题搁在一边——牛顿的有些追随者会承认,这只有在造物主的直接命令下才会发生——牛顿证明了一个重要的定理,合并的拉力等于设想整个地球全集中于地心的拉力。

他现在能够将地球对石头的拉力(在 1 个地球半径的距离上有效)和地球将月球送入封闭轨道(距离为 60 倍地球半径)的拉力相比较。他发觉拉力之比的确为 $60^2: 1$。地球和天空符合引力的反平方定律。

当《原理》的文稿增加时,能找到解释的现象的数目也在增加。潮汐源自太阳和月球对陆地和海洋的引力效应之间的差别。自转的地球在赤道隆起而在两极平坦,因而不是严格地呈球形;结果是,太阳和月球的引力造成了地轴的摆动,并且由此引起了喜帕恰斯首先注意到的分点的岁差。月球运动中的几个"不等量"或不规则性也被发现——一个由托勒密发现,另一个由第谷发现——这些也是牛顿能定性地(即使不是定量地)作出说明的。

我们的卫星是易于观测的,虽然很容易说它受到了引力,但从数学上分析则是高度复杂。牛顿为 18 世纪最具才能的数学家设置了一项任务:证明观测到的月球运动完全能够用反平方定律来说明。胆小鬼不敢去对这些逐步走向成功的尝试作历史性研究;对我们来说幸运的是,与其说它属于天文学史,倒不如说它属于应用数学史。

牛顿能够用观测到的地球、木星和土星的卫星去计算它们母行星的质量,他发现木星和土星要比地球巨大——而且,十之八九要比水星、金星和火星大。因此,这两颗大质量行星似乎位于太阳系的外围,在那里它们的巨大引力对太阳系稳定性的伤害会最小。但是即使是这种天意的安排也会经受摄动,然后系统会"需要一种改革":天意会进行干涉来恢复原始的秩序,从而证明上帝对人类的持续关怀。

有些欧洲大陆的学者,如著名的莱布尼茨(1646—1716),认同牛顿的观点,认为上帝是伟大的钟表师,并且将宇宙视为机械装置的一个杰作,但是对于牛顿把上帝想作是一个可怜的、得用创造奇迹这种方式以修正其大错的工人而感到大为愤慨。但是,对于牛顿来说,从一开始这就是上帝计划的全部;为展现对其创造的继续眷顾,上帝已经参与了与宇宙的一个服务合约。

其余的欧洲大陆人发现引力的概念是个倒退:牛顿的确用这个设想的力解释了许多运动,但引力本身到目前为止还是不可解释的。它能用笛卡尔原理来说明吗?牛顿《原理》中定理的陈旧几何表述足以吓住所有人,除了少数自以为是的读者以外。直到 18 世纪早先几十年里推广普及开始出现,特别是大陆数学家成功地利用牛顿的方案并且解释了月球复杂运动中越来越多的方面时,引力的价值才变得无可置疑。1759 年一颗彗星重现时,任何残留的疑虑都被一扫而光。

按照笛卡尔的物理学,一颗彗星是一颗死了的恒星,它自己的漩涡已经破灭,然后它从一个漩涡飘到另一个漩涡,如果它穿入一个漩涡足够深的话,它仍然可以以行星的样子待在那里。但是牛顿声称彗星符合开普勒定律(以其推广了的形式),而且轨道为扁长椭圆的彗星会有规律地重现。因此哈雷搜寻历史记录,寻找 3 颗或更多具有相似轨道特征的彗星,它们的再现在时间上间隔同样的年份或其倍数；他发现 1531 年、1607 年和 1682 年的彗星看来吻合这一规律。在 1695 年，他告诉牛顿，他想这些是同一颗彗星的重现。

但是，它们的时间间隔虽然相似，却并非相等。哈雷意识到，这是因为轨道会发生更改，每当彗星穿越太阳系时，彗星会在大行星近旁经过，并且经受该行星的引力拉动；他预报同一彗星将于"大约 1758 年的年末或次年的年初"回归。

这些奇特的初现能够像行星那样有规律吗？在 1757 年的夏天，克雷洛（1713—1765）和他的两个同事借助于钟表较详细地计算出了这颗彗星的轨道在 1682 年如何变更，当时它离开太阳系在木星近旁通过；最后，他们能够预测这颗回归的彗星在 1759 年 4 月中旬的几周内在太阳附近转向。

1758 年的圣诞节，人们的确看到了一颗新到达的彗星，它在 1759 年 3 月 13 日绕太阳环行。关键的一点是，它的轨道特征与哈雷已经研究过的 3 颗彗星的轨道特征非常相似：所有 4 颗彗星是同一颗彗星。使天文学家和公众惊讶的是，牛顿力学预报了"哈雷彗星"在间隔了 3/4 个世纪后的回归。

同时，对于月球复杂行为的分析花费了许多数学努力。这部分是由数学上的好奇心所推动的，但也有多得多的严肃目的。海上水手的生命取决于他们知道他们在哪里，尤其是在夜间。确定船舶的纬度是比较直接的：领航员在夜间测量天极的地平高度（或者，不那么直接地，在中午时测量太阳的地平高度）。确定经度——对今日的航空旅行者来说，时间改变太熟悉了——则要困难得多。人们该如何比较地方时和标准时（今天我们用的是格林尼治标准时间）呢？18 世纪早期，摆钟在陆地上走得还比较准，但到了海上就没有用了。

几个世纪以来,不时有人采纳喜帕恰斯的建议,认为城市之间的经度差,可以由两个地方同时观测月蚀,比较其地方时来确定;但是这样的食象对于航海者来说实在是太罕有了。伽利略提议用寻常得多的木星卫星的交食来代替;到了17世纪晚期,精确的木卫表使得这一方法在陆地上得到了成功的使用。但这样的食象——公平地说,依然是很罕见——几乎不可能从船上观测到。

有人也曾尝试过另外的方法,既有几近无用的方法,也有离奇古怪的方法,最后严肃的选择方案归结为两个:研制能在海上维持精确时间的天文钟,以及利用月球相对于恒星的快速运动(与钟表时针相对于日晷时钟示数类似)。英国议会为海上经度问题的实际解决方案提供的奖金将使领奖人一夜暴富。

制作天文钟是手工钟表匠的活儿,其中为首的是约翰·哈里森(1693—1776)。同时,大学培养出的天文学家和数学家则为完善“月球距离”的测量方法而努力。为将这个方法付诸实施,领航员必须首先确定出月球在天空中的现时位置——实际上,是它相对于附近恒星的位置。为此,他需要一张准确的恒星星表以及一架用来量度月球和附近恒星之间角度的精确仪器。然后他要求有可靠的月球用表,将月球的观测位置转换成标准时,用来与他的地方时作比较,从而给出他的经度。恒星位置、角度测量以及月球用表的误差都会加大船舶实际位置和领航员计算出的船舶位置之间的差距,因而,将三项误差中的每一项尽可能地减少是很重要的。

位于格林尼治的皇家天文台于 1675 年创立,专门为了满足航海者对一张精确的恒星星表的需求;1725 年发表的弗拉姆斯蒂德的遗著《不列颠星表》中含有 3000 颗恒星,它比第谷的肉眼恒星星表改进了整整一个星等,它的出版是第一个皇家天文学家对此需求的满足。适用于海上测量角度的精确仪器于1731年问世,这是一台双反射象限仪(六分仪的祖宗)。该由数学家(实际都是法国人和德国人)来完善牛顿的月球理论了,他们要整理出一张月球位置的精确用表,能够提前数月计算出月球的位置供航海者使用。终于,格丁根的教授图比亚斯·迈耶尔(1723—1762)研制出了表格,好得足以使他的遗孀挣得英国提供的3000英镑的奖金。所有这些使得当时的皇家天文学家奈维尔·马斯基林(1732—1811)能够在1766年出版《航海历书》的首本年刊。

与此同时,哈里森正在制作一个又一个巧妙的天文钟。第一个被送到里斯本作试验用的于1736年被运回。结果是振奋人心的,哈里森因此获得250英镑,以资助其作进一步的研究和发展。就这样大约30年之后,1764年哈里森带着他第四个天文钟航行到了巴巴多斯而后返航,事后他被授予原先作为奖金提供的20000英镑的一半。一旦合适的天文钟能够以批量生产,它们就变成了经度问题更可取的解决方案。天文学家发觉他们自己还有一种新的作用,给大港口的天文台配备人员并在正午(或午后一小时)投放报时球,使得航海员在起航之前能够检查他们的天文钟。哈里森的天文钟——运动中的诗——今天能在格林尼治国立海事博物馆里看到。

牛顿看到了将小的内行星与大质量的木星和土星分隔开的巨大间隙,并视之为上帝保护太阳系免遭瓦解的证据。但是开普勒早先提出过这样的概念:这个间隙被一颗迄今尚未发现的行星所占据。到了18世纪,"已知行星"(言外之意是或许还有别的行星仍属未知)这一说法已不再是不寻常的了,一个奇妙的依行星离日距离排位的算术数列的发现使得人们纷纷推测可能有一颗"失踪的"行星存在。在 1702 年出版的《天文学基础》中,牛津教授戴维·格里高利(1659—1708)将这些距离定为正比于 4、7、10、15、52 和 95;将其中两个数字稍作变更后,维滕贝格的约翰·丹尼尔·提丢斯(1729—1796)让它们等于 4、4+3、4+6、4+12、4+48 和 4+96。这些数字具有($4+3 \times 2^n$)的形式。这个算式被年轻的德国天文学家约翰·艾勒托·波得(1747—1826)热情地采用了,今日称之为波得定则。提丢斯和波得同意,定然有或曾经有对应于项 $4+3 \times 2^3$ 的一个或数个天体。

　　1781 年,一桩完全意想不到的事情发生了:一个业余观测者威廉·赫歇尔(1738—1822)——关于他我们将有很多话要说——在研究较亮恒星时,偶然见到了一个奇妙的天体,它被证明是颗行星,今日我们称之为天王星。当数学家能够定出其轨道时,他们作出了重大发现,认为其与太阳的距离符合下面这个数列的下一项:$4+3 \times 2^6$。这足以使哥达的宫廷天文学家冯·察赫(1754—1832)对数列样式的确实性感到信服,他开始搜寻对应于项 $4+3 \times 2^3$ 的一颗行星。在未获成功之后,1800年,他和一群朋友举行了一次会议,讨论怎样最好地继续下去。他们将黄道带——任何行星可能存在的区域——划分成 24 个区;每个区指定一个特别的观测者,他的职责是管辖他的区并且寻求没有固定驻留地的任何一颗"恒星"。

　　西西里巴勒莫天文台的朱塞佩·皮亚齐(1746—1826)是他们看中的巡视员之一。当时皮亚齐正在研制一张恒星星表,他工作得很仔细,在继后的夜晚会重新测量每颗恒星的位置。1801 年 1 月 1 日晚,在冯·察赫和同伴的邀请到达之前,皮亚齐正在测量一颗八等"恒星"的位置,当他重测这颗星时,他发觉它已经移动了。

在太阳的炫光中丢掉它之前,皮亚齐也只能跟踪这个天体几个星期。在该年年末,冯·察赫又重新发现了它,这得感谢新出现的数学天才卡尔·弗里德里希·高斯(1777—1855)。皮亚齐称呼它为谷神星,它与波得定则的缺失项正好相符,但是它是小个儿的:赫歇尔(正确地)认为它甚至比月球还要小。还要糟的是,还有三个这样的天体,也是既小又符合缺失项,也在接下来的几年中被发现。赫歇尔提出将这些天体新族的成员称为"小行星"。内科医生和天文学家威廉·奥伯斯(1758—1840)也卷入了对失踪行星的搜寻,他认为它们可能是曾经存在过的某颗行星的碎片。

搜寻持续了许多年,但是没有结果,最后也就放弃了。直到1845年,另一颗小行星被德国的前邮递员亨克所发现。他的第二个成功发现是在两年之后,这一次重新燃起了大家的兴趣。到了1891年,被找到的小行星已经多于300颗。照相术现在已经简化了搜寻。海德堡的马克斯·沃尔夫(1863—1932)能够用一架追踪天空移动的望远镜在几个小时内拍摄一个大星场;恒星将以光点形式出现,小行星相对于恒星运动时会留下一条光迹。

假如奥伯斯是正确的,那么每颗小行星的轨道会——至少初始会——通过行星瓦解之处和太阳反面的对应之处。事实证明情况不是这样的。天文学家现在认为小行星合并起来的质量只是月球质量的很小一部分,由于木星的吸引力,它们是不可能聚合成一颗行星的。

天王星的发现延伸了波得定则的序列,但很快就发现,这颗行星的运动是令人费解的。其轨道的早期确定因为发现它早在 1690 年就被观测（并且被列为一颗恒星）而被大大地简化了，但是该行星随即偏离其预定路径。各种各样的解释被提了出来，最后缩减为两个——或是反平方定律的公式在这样的距离上需要修正，或是天王星被一颗至今仍未发现的行星所拉动——并且归结为一个：未发现的行星。到了 19 世纪 40 年代，两位天才数学家伏案工作，用笔在纸上计算，希望告诉天文学家何处去寻觅这颗未知的（也是先前意料之外的）太阳卫星。

两个人中年纪小一些的是剑桥的研究生亚当斯（1819—1892）。詹姆斯·查理斯（1803—1882）是他的教授。1845 年秋天，在查理斯的建议之下，亚当斯访问了格林尼治，向皇家天文学家埃里（1801—1892）解释他的计算。由于运气不好，他未能亲自见到埃里，但他留下了结果的摘要。次年夏天，埃里惊讶地收到了巴黎勒威耶（1811—1877）论文的副本，该论文预言了一颗几乎总处于同一位置的行星的存在。埃里的见解是，那种研究不是他所管理的国家天文台的研究范围，但他要求查理斯在剑桥展开搜索。

查理斯只好小心翼翼地在某一天区中标绘出星状天体，然后在一天后回到同一区域看看它们之中是否有一个移动了。这不可避免地是一个令人腻烦而且耗时的过程，查理斯也并不着急。不幸的是，为了英法关系的未来，勒威耶已要求柏林天文台的天文学家作搜索。他们——不像查理斯——拥有柏林科学院新星图有关图片的副本，所以能够将天空中的恒星与星图中的恒星进行比对。在 1846 年 9 月 23 日开始搜寻的几分钟以内，他们就找到了一个不在星图上的星状天体。它正是那颗失踪的行星。

后来人们知道,查理斯曾注意过同一颗"恒星",但是他还没有回到同一天区里重新测量其位置。对英国人来说,亚当斯对行星海王星发现的道义申索,等于勒威耶的道义申索,但法国人却不是这么看的。但是,不管这个优先权的问题争论得如何,牛顿力学的地位则得到了充分肯定。

　　这一令人愉快的情形并没有持续。像天王星那样,水星的轨道有一个无法解释的特征:它最靠近太阳的点在经度上的进动比预料中的更快,大约每一个世纪超前约 1 度——虽然不多,但仍然要求解释。勒威耶不禁怀疑还有另一颗看不见的行星;1859 年 9 月,他宣告说,一颗大小和水星一样、离日距离只达一半(从而难以观测)的行星可能是该现象的一种解释。碰巧,一个名叫莱卡博的不知名的法国内科医生在该年初曾经看见一个天体横穿太阳(或者他是这么认为的),当他读到勒威耶的预报后,他写信给勒威耶。勒威耶说莱卡博的观测是可靠的,他很满意,并将内科医生曾经见过的这颗行星命名为祝融星。各式各样声称见到了祝融星的断言接踵而至,但很少能令人信服;到了本世纪末,祝融星被认作是虚假的而终被舍弃。1915年,爱因斯坦指出,水星的异常行为正是他的广义相对论所隐含的:对宇宙来说,还有许多是超出牛顿哲学想象的。

第六章
# 探索恒星宇宙

　　直到 1572 年,天王星家才把"固定的"恒星——不只是在位置上,而且在亮度上都是固定不变的——视为行星运动的背景。当然,事实上恒星具有越过天空的独特或"固有"的运动,但是星际距离是如此之大,以至于即使是发自最近恒星的光也要几年后才会到达地球。结果,自行几乎是不可察觉的,除非在很长的时标上;因而,文艺复兴时期观测者所见的恒星的位置与托勒密赋予它们的位置没有什么不同(除了岁差的总效应之外)。

　　没有注意到亮度的变化可能是更令人惊讶的。虽然大多数恒星,比如太阳,几乎是不变的,少数恒星被一个伴星交食后亮度衰减,其他恒星则会经受较大的物理改变,无论是规则的还是不规则的,从而发生亮度变化。但是这些变星中没有一颗能亮到使其亮度变化让中世纪观测者相信天区是变化的。当你已经知道改变已是不可能时,为什么要寻找改变呢?

　　大自然唤醒了我们,1572 年新星的出现(正如第谷所见,参见第四章)使我们相信了恒星是会变的,从而燃起了我们对它的兴趣。另一颗这样的新星则在 1604 年爆发,它在欧洲造成了惊恐和沮丧。8 个世纪以来的第一次,缓慢运动的行星木星和土星在黄道带致命的"火焰区"相合;新星在它们中间闪耀,火星也参与其中——能想象到的最不祥的占星事件。

现在没有人怀疑天空会发生变化。的确,还有人谈到出现于鲸鱼座的另一颗新星,但是它要暗一些,在它暗淡下去和消失之前,只有一个观测者见到过它。1638 年,鲸鱼座成为第二颗新星的宿主(或者说,看起来是如此);像它的前任那样,它逐渐变暗并且消失——但是在它的发现者能够发表他的描述之前,它又重现了,这使他很惊讶。它继续每隔一段时间就消失和重现,1667 年,伊斯梅尔·布里奥(1605—1694)宣布这颗"奇妙的星"每隔 11 个月达到其最大亮度;它的运动在某种程度上是可以预报的,因而是有规律的。

布里奥继续给出了变星的一个物理解释,那是一个非常巧妙的解释。他指出,太阳黑子的变化表明,太阳本身——此时仅仅被认为是最靠近我们的恒星——严格说来是变化的。而且太阳黑子的转动证明了太阳整体是自转的,其他恒星无疑亦是如此。然后设想一颗布满黑斑而不仅仅是太阳黑子的自转恒星;每当一块黑斑朝向我们时,我们将看到恒星的亮度减弱,这将随着恒星的每一次自转而有规律地发生。但若黑斑本身像黑子那样不规则地变化,则将造成亮度的不规则变化。用这种方式,布里奥能够说明规则的和不规则的变化。的确,他成功地从物理学角度解释了变星,天文学家心满意足地宣布了他们关于特殊恒星变化的发现。但是,这些宣布既不易被证实又不易被证伪;被说成是变星的天体的数目激增,整个论题可说是声名扫地。

到了 18 世纪末,当威廉·赫歇尔发表了一系列《恒星的比较亮度表》时,核实这样的宣布的任务就被简化了。在这些表中，赫歇尔将恒星与其附近具有相似亮度的恒星仔细地比对，如此一来，其中一颗恒星的变化将会扰动已发表的比较星从而显露其自身。赫歇尔推广了一种方法，按亮度等级排列恒星的序列，这个方法在 18 世纪 80 年代早期被两位业余天文学爱好者所发展，他们是英格兰北部约克市的一对邻居。爱德华·皮戈特（1753—1825）是一位有造诣的天文观测家的儿子；他的年轻朋友约翰·古德里克（1764—1786）是个聋哑人，他热情地接受了邀请，参与了变星的研究。

他们详细研究的恒星之一是大陵五，一个世纪之前，它曾有两次被报告为四等星，而不是通常的二等星。1782 年 11 月 7 日，大陵五仍是二等星，但 5 天后，它减弱为四等，次夜，又变回到二等。如此速率的变化前所未有，所以这两人一直监视这颗恒星。12 月 28 日，他们的努力得到了回报：黄昏时他们见到的大陵五为三等或四等星，但就在他们眼前，它增亮为了二等。皮戈特马上怀疑大陵五正被一颗卫星所食，次日他向古德里克送出一份报告，在那里面，他假设在 11 月 12 日至 12 月 28 日间的 46 天里，卫星完成了一个或两个轨道，并由此计算出了这颗假想中的卫星未来的轨道。事实上，在即将到来的几个月中，他们的观测表明，卫星——如果的确是这种解释——绕大陵五运转一周不足 3 天。对天文学来说，这是一个迄今未知的现象。

皮戈特慷慨地将向英国皇家学会正式报告之事留给了他的残疾朋友，但十多岁的古德里克只是说交食理论为可能的原因之一，和传统的黑斑说一样。皮戈特事实上是对的——大陵五的确是被一颗伴星所食——但是这两位朋友最终又回到了黑斑解释，可能是因为他们错误地认为他们看到了大陵五光变曲线中的不规则性，或是因为他们发现的另外 3 颗短周期变星不能用交食理论说明。事实上，其中两颗是造父变星，即迅速升至最大亮度而后缓慢变暗的脉动恒星，有一天它们会被埃德温·哈勃和他的同时代人用作距离指示物。

结果是 18 世纪天文学贡献出了一族新的变星，它们的周期只有几天，但在理解这些现象的物理原因方面，进展甚微。

皮戈特和古德里克曾见到大陵五的亮度在几小时内发生变化。相比之下，位置的变化则要在很长时间后才能观测到。相对而言，只有很少的恒星具有高达每年 1 弧秒的自行，已知最大的自行也只是刚刚超过 10 弧秒。这样的运动只有在比较了恒星的现在位置与记录在星表中的早先位置后才能发觉；在其他条件相同时，距离早期星表的时间间隔越长，得到的自行值就越精确。但不幸的是，其他条件并非相同；当我们沿时间回溯时，精度标准下降，早期星表中任何不精确的恒星位置将会影响所得出的自行精度。

古代唯一的星表在托勒密的《天文学大成》中；1718 年爱德蒙·哈雷用这张星表定出黄道倾角——黄道对赤道倾斜的角度——的变率时，认识到 3 颗恒星定然是相互独立地运动着。

此时哈雷进一步研究这个问题就不容易了，因为过去唯一一张有价值的星表就是第谷的星表了。它比托勒密的星表要精确得多；但是它距此时只有一个世纪多一点，并且它的作者只是粗糙地处理折射，即星光进入地球大气后的弯曲（它影响了恒星在天空中的观测位置）。不过，未来的几代人能够用上约翰·弗拉姆斯蒂德在格林尼治精心编制的《不列颠星表》作为时间的起点去量度恒星的自行了。

或许只是看起来如此。但是后来在 1728 年，詹姆斯·布拉德雷（1693—1762）宣布了一个完全出人意料的复杂情况：

"光行差"。光的速度很大,但是,正如上世纪晚期对木星卫星交食的观测所表明的,它仍然是有限的。当木星靠近地球,携有交食信息的光不必走那么远时,交食会提前发生。当木星离开太阳背面时,交食会推迟发生。

比较起来,地球在其绕日轨道上的速度是小的,但它又大到足以影响恒星的观测位置。在观测者看来,一颗恒星位于星光到达的方向上;这个方向随着地球运动方向的改变而(微微地)改变——如同事实上垂直落下的雨似乎从我们向之运动的方向上打在我们的脸上一样。

我们将在本章的后面看到布拉德雷是怎样发现了光行差。他的发现暗示,即使是用《不列颠星表》作自行测量的时间起点也是有严重缺陷的。当布拉德雷在 1748 年宣告地球的轴有"章动"或摆动时,该星表的另一个欠缺也暴露了出来。这是因为地球不是一个理想的球体, 太阳和月球对地球的引力拉动有变化,并且这也引起了用于测量恒星位置的坐标系的运动。

布拉德雷本人在 1742 年成为了皇家天文学家。从 1750 年起直到他的健康开始衰退为止,他执行着一项观测计划,在其中他谨小慎微地记录着能够影响恒星观测位置的所有情况。但是他自己没有能来得及"归算"他的观测——作出为导出他的恒星的真正位置所需的计算。归算直到 1818 年才完成,当时伟大的德国数学家弗里德里希·威尔海尔姆·贝塞尔(1784—1846)出版了贴切命名的《天文学基础》,该书包含了 1755 年超过 3000 颗恒星的位置, 那时正是布拉德雷观测计划中一个方便的时间。从那以后,19 世纪的天文学家能够将一颗恒星的现时位置与《基础》中给出的该恒星在 1755 年的位置作比较,从而确定每年运动着的恒星在这段时间间隔中穿越天空有多远。

　　布拉德雷自己在 1748 年指出,所有的自行都是相对的:我们并没有观测到一颗恒星在绝对空间中怎样运动,而只是观测到它相对于我们如何运动。12 年之后,图比亚斯·迈耶尔讨论了这一点的含义。如果除了太阳以外,每颗恒星都是静止的,则太阳系通过空间的运动会向我们揭示其在恒星之间的(视)运动样式。所以,已知自行中的任一样式可能反映了太阳系的一种运动;剩余运动将会是单颗恒星本身的剩余运动。

　　一个现代的类比阐明了这样一种样式是怎样的。如果在城市里夜间驾驶一辆小汽车,一簇远处的车灯似乎合成一束,但当我们靠近时,它们似乎又分开了。同时,我们左边的街灯似乎在反时针运动,而我们右边的街灯似乎在顺时针运动。

　　迈耶尔没有在他所知的(不可靠的)自行中找到这样的样式,但在 1783 年,威廉·赫歇尔——有一段时间完全扑在他的书桌上工作——相信他已经找到了一种样式,这种样式意味着太阳系正向着武仙座运动。今天,无人怀疑他的结论,但是他的论据经不起仔细的调查。一代人以后,贝塞尔发现了找出可靠自行的方法,当时的那几个月他的《基础》正在印刷,并且他拥有了阐明任何一种运动样式所需的全部数学才能,但是他只留下了一片空白。

　　到了 1837 年,天文学家才相信一种解决方案就在眼前。在那一年,波恩的天文学教授阿格兰德尔(1799—1875)发表了不少于 390 个自行的分析。他将自行按大小划分成 3 组,每组独立地给出了一个太阳向点的方向,离赫歇尔提出的方向不远。

　　他的结论很快被其他天文学家的分析所证实,但是这些全都依赖于同一基本资料——布拉德雷在英格兰对恒星所作的观测。但是拉卡伊(1713—1762)在 1751—1753 年造访了好望角并且定出了差不多 10000 颗恒星的位置,这些南天恒星中有部分在 19 世纪的位置当时也已经知道。1847 年,保险统计员托马斯·加罗威(1796—1851)分析了 81 个自行(它们同布拉德雷完全无关)并且导出了一个方向,与基于北天恒星资料导出的方向相似。此后没有人再怀疑太阳系正按武仙座方向运动,对于已知自行——随着时间推移而在数量和精度上大大增加的一览表——的进一步分析只被用来完善这个理论。

恒星有多远?对于古时的托勒密和 16 世纪晚期的第谷·布拉赫来说,固定的恒星只是在最外的行星之外。但若哥白尼是对的,则每 6 个月,我们就能从长度为地球绕日半径两倍(两个"天文单位")的巨大基线的相反两端观测恒星。正如我们已经看到的那样,即使第谷用他的精密仪器也不能检测出恒星之间会产生的视运动("周年视差"),他很合理地将此视为了对日心假设的驳斥。

问题部分出在观测的性质:当月份过去时,温度和湿度的季节变化将造成仪器的翘曲,空气压力的改变会使折射发生变动等。伽利略像往常一样机灵,他想到了一个办法来克服这些困难。假设两颗恒星相对于地球位于几乎同一方向,并且假设一颗比另一颗要远得多。

较远的恒星要比较近者有小得多的视差;这意味着,如果我们全然忽略较远恒星的视差并且取它为天空中的一个准固定点,再由它来测量较近恒星的视差,我们也不会错得太离谱。但这样做的便利非常明显,因为这两颗恒星将因仪器的任一翘曲、折射的改变等等受到相等的影响,可将这样的复杂效应从考虑中剔除。

懒惰一如往常的伽利略并没有证实他自己的观点,好多年后才取得了进展。与此同时，勒内·笛卡尔让学界相信，恒星是太阳，而太阳只是我们区域内的恒星。这是对恒星距离问题的一种新观点。

倘若空间是完全透明的，光就会按照距离的平方律衰减。所以，如果太阳被移至比它现在远上 1000 倍处，它的亮度将只

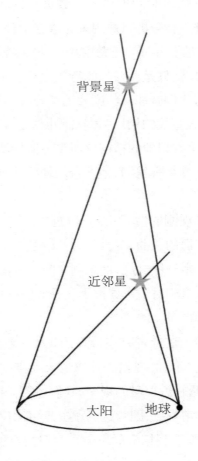

图 15　伽利略检测周年视差的方法，测量一颗近邻恒星相对于一颗背景恒星的周年视运动。

有其现有亮度的百万分之一。现在假定恒星不止性质上相似于太阳而且在物理上恒同于太阳,这样从物理上来说,天狼星(还有其他恒星)是太阳的双胞胎兄弟。于是如果天狼星的亮度是太阳亮度的百万分之一,假定空间是透明的,那么我们就会知道天狼星要比太阳远上 1000 倍。

但是,人们怎样在明亮太阳的辉光和恒星的暗光之间作比较呢?荷兰物理学家克里斯蒂安·惠更斯(1629—1695)在他自己和太阳之间放置一屏幕,其上钻一小孔。他的意图是改变孔的大小,直至穿过小孔的这部分可见光在亮度上与天狼星相等,然后计算太阳的什么部分是可见的。这是一种粗略的办法,但是他的结果——天狼星距离我们 27664 天文单位——是 1698 年被发表后超过四分之一个世纪中被印出来的唯一估算,因而被广泛地引用。明显地,恒星离我们很远。

与此同时,除了他的小圈子以外所有人都不知道的是,艾萨克·牛顿采用苏格兰数学家詹姆斯·格里高利(1638—1675)的独创性建议已经取得了好得多的进展。在一本出版于 1668 年的很少有人注意的书中,格里高利提出用一颗行星代替天狼星来简化亮度比较。直到该行星在亮度上与天狼星相等,人们才利用太阳系内的尺度知识,将直接照射到地球的太阳光与通过该行星反射到地球的太阳光相比较。按照这一思路,牛顿将天狼星放在 1000000 天文单位处。碰巧,天狼星比那个距离的一半还要远一些,故而,牛顿的熟人此时完全意识到了太阳与邻近恒星之间距离之大。

但是,每颗恒星是太阳的一个双胞胎兄弟这种暂时的假设是无法替代对于特定恒星周年视差(和由此得出的距离)的实际测量的。罗伯特·胡克想到,因为天龙座 γ 从他伦敦寓所的头顶通过,所以它的星光不受大气折射的影响。他将望远镜的部件与房子的实际构造结合起来,尝试着规避观测仪器季节性翘曲的危险。虽然望远镜天文学尚处于其初级阶段,但胡克设计并建造了一架望远镜,只为了在恒星通路上的某一时刻作观测。

胡克虽然足智多谋但却没有坚持不懈:1669 年他的疾病和一个望远镜透镜事故使他停止了努力,他仅仅作了 4 次观测。但是他的方法有许多可以推荐之处。18 世纪 20 年代中期,一个富裕的英国业余爱好者塞缪尔·莫利纽克斯(1689—1728)决定作另一次尝试,去测量天龙座 γ 的周年视差。他邀请詹姆斯·布拉德雷参与了他的工作,并且委托一个杰出的制造商乔治·格雷厄姆制作了一具"天顶扇形仪"。这架有垂直望远镜的仪器被安装在了莫利纽克斯家的烟囱上,当恒星从头顶越过时,望远镜的镜筒稍微倾斜,以使恒星从视场中央通过。倾斜的角度可以按天顶扇形仪的刻度测量出来,以给出恒星相对于垂线的角距离。

简单的计算表明,天龙座 γ 应当在圣诞前一星期到达一个极南的位置,所以布拉德雷觉得奇怪,他在 12 月 21 日看到它从头顶通过,明显比一星期前位置更靠南。到了次年 3 月,它应当向北移动时,它走到了比其 12 月时的位置更靠南大约 20 弧秒处。然后这颗恒星停下来走回头路,到 6 月回到它在头年 12 月时的位置并且在 9 月到达其最北端。

莫利纽克斯和布拉德雷这两位朋友争论着各种解释——是否有一种地轴的运动从而导致了我们用以测量恒星位置的坐标系的运动?或者地球大气因行星通过空间遭到了畸变,使得大气折射意想不到地影响到了测量?——但是没有结果。布拉德雷委托格雷厄姆制作了另一架天顶扇形仪,这次有了更广阔的视场,能观测到更多的恒星,因此他建立了恒星运动的样式;但是,它们的解释使他困惑。后来有一天,在泰晤士河的一条船上,他注意到当船转向时,船上的风向标也随之转向——当然不是因为风改变了方向,而是因为船改变了航程。他现在意识到星光同样是从改变着的方向抵达观测者,因为当地球环绕太阳运转时,观测者也改变着位置。

1729 年向皇家学会宣布的光行差的发现是重大的,其中有几大理由。它是地球绕日运动的第一个直接证据。因为所有恒星受到相似的影响,所以这说明光的速度是自然界的一个常数。它揭示了(如同我们早先看到的)在过去的恒星位置测量(包括弗拉姆斯蒂德的测量)中有一个完全意想不到的错误。连精度确切的布拉德雷的天顶扇形仪也不能观测到周年视差,因此恒星必定位于至少 400000 天文单位处。

就在前一年,牛顿遗作《宇宙体系》的出版公开了他的估计。的确,这基于恒星之间的物理一致性这一工作假设——认为天狼星位于百万天文单位处。这两个结果——一个给出实际距离但是立足于一个有问题的假设,另一个则给出最小距离但是基于直接测量——合在一起使天文学家相信,恒星距离的尺度终被了解了。

其中隐含着一个不受欢迎的结论:周年视差最多为 1 或 2 弧秒,这个角度是那么小,相当于几公里外一枚硬币的宽度。在几个月内进行的这样一个一分钟的运动几乎不可能被观测到,下一代的天文学家对这样一个无望的任务没有表示出多少热情。威廉·赫歇尔在 18 世纪七八十年代收集了大量双星,表面上是为了用伽利略的方法测定视差;但他像自然史学者那样,收集了旁人有一天可能要用到的标本。大多数天文学家倾向于将他们的时间花在更有希望的调查方向上。

总之，约翰·米歇尔（约 1724—1793）曾在 1767 年指出——赫歇尔并不知道——双星的数目很大，其中大多数必定是空间里真正的成对者（双子星），它们与观测者的距离相同，因而对测定视差的伽利略的方法是没有用的。当赫歇尔在世纪之交重新检查他的某些双星时，他自己证实了米歇尔的观点，而且还找到了两颗恒星相互运转的例子。一代之后，威廉的儿子约翰证实了它们的轨道是椭圆，并且将这些伙伴星结合在一起的力是牛顿的引力（他并不是唯一持这一观点的人）。虽然牛顿曾经宣布引力是一种普适的定律，但是这是定律应用到太阳系之外的第一个证据。

同时，天文学家发觉他们自己处于这样一种境地：随着望远镜的改进，天球上恒星位置的两个坐标以日益增加的精度被测量出来，而对恒星的第三个坐标——距离——则除了其巨大的尺度以外，知之甚少。当已知自行的数目增加，人们发现了并非所有快速运动的恒星都是亮星时，就连最近的星最亮这个假设也成了问题。

一个极端的例子在 19 世纪早期就被找到了。先是皮亚齐然后是贝塞尔发现，相对较暗的恒星天鹅座 61 以每年超过 5 弧秒的罕见速度穿越天空。这是否一定表明这颗恒星尽管其亮度不高但必定距离我们很近？

周年视差当然是与距离成反比的。试图测量视差的观测者应将他们的努力集中在最靠近地球的恒星上，这是很重要的。1837 年，在数次宣布测量成功，而后又被证明站不住脚之后，德国出生的威廉·斯特鲁维（1793—1864）提出了近距离恒星的三个判据:这颗恒星是否很亮?其自行是否很大?如果是一对双子星的话,考虑到两颗子星相互运转的时间,这两颗子星是否看上去彼此分得很开?

在多尔帕特(今为爱沙尼亚的塔图)的天文台里,斯特鲁维被特许拥有一架由约瑟夫·夫琅和费(1787—1826)赠送的大型折射望远镜,它的物镜玻璃直径不小于 24 厘米且质量非常好,并且它的安装方式和赤道相似,它的轴指向北天极,所以观测者只需转动一个轴,就可以使望远镜和恒星排成一行。为测量视差,斯特鲁维选定了织女星,它很亮,具有大的自行。1837 年他宣布了 17 次观测的结果,由此推论出视差为 1/8 弧秒。三年之后,他报告了 100 次观测,这次推断出视差为 1/4 弧秒。但是,自胡克起,虚假宣布就一直不少,所以天文学家仍免不了心存疑虑。

其时,柯尼斯堡的贝塞尔同样幸运地获得了很好的仪器。他的夫琅和费折射镜并不是大型的,物镜直径只有 16 厘米。但是它的制作者不满足于得到一个高质量的透镜,他勇敢地将它分割成两块半圆形的玻璃片,它们能够沿着公共直径相互运动。每个半圆都有一个完整的像,而像的亮度则只有原来的一半。如果望远镜转向一对双星,它们会出现在每个半圆上,观测者能够将一个半圆相对于另一个半圆滑动,直至一个像上的一颗恒星与另一个像上的另一颗恒星恰好相合。需要的位移非常精确地指示了分割两星的角度。因为这样的仪器常用于监视太阳视直径的变化,所以它们被称为量日仪。

贝塞尔选择了被称为"飞星"的天鹅座 61 作仔细观测,因为其自行很大。1837 年,他让这颗恒星受到前所未有的观测,每个晚上观测 16 次,在"能见度"特好的情形下,观测次数还要更多,就这样观测了一年多。次年他就能够宣布这颗恒星的视差约为 1/3 弧秒。具有说服力的是,由他的观测绘制的图与预料中的理论曲线相吻合。约翰·赫歇尔告诉皇家天文学会,这是"实用天文学曾目睹的最伟大、最辉煌的胜利"。恒星的宇宙现在有了第三维,成功测量的周年视差的数目在未来的几十年中会成倍地增长。

但是这宇宙的大尺度结构是什么?牛顿的《原理》几乎没有谈及恒星。1692 年在收到年轻的神学家理查德·本特利(1662—1742)的一封信之前,牛顿对宇宙学问题也没有什么想法。本特利曾就科学和宗教作过一系列的讲座和布道,在将这些付印以前,他想知道那本人人尊敬但却无人能读懂的浓缩的数学书作者的观点。本特利没有时间来研究笛卡尔的立场,笛卡尔认为上帝创造了宇宙,并放手让其自行运动;但是他想知道,这一观点的论据是什么,于是他问牛顿,在一个初始时物质呈现理想对称的宇宙中会发生什么。牛顿没有意识到本特利指的对称是完全理想的,他回答说,在任何地方物质若比寻常更加稠密,其引力将会吸引周围的物质并导致更大的密度。本特利说牛顿说得不对,这使牛顿大为恼火,继而他承认在一个**理想**对称的宇宙里,物质没有理由以一种方式而不是另一种方式运动;但是他评论说,理想对称是有问题的,正如无穷多的缝衣针全都针头朝下立在一个无限大的镜面上。"这不是一样难吗?"本特利反驳道,"在无限空间中无限多这样的物质也要维持平衡呢。"换句话说,当每颗恒星都被所有其余恒星的引力拉着时,恒星是"固定"不动的,这不是一样吗?

《原理》声称引力是自然的普适定律,现在牛顿正面临着矛盾;因为即使经过多个世纪的观测,恒星似乎仍像以前一样固定。稀奇的是,牛顿是唯一一个(正如我们已经看到的)对星际距离的尺度有正确估计的人;但是他没有想到,恒星是那么遥远,它们的任何运动几乎都是不可察觉的。他继续相信恒星是不动的,他的问题是要解释何以会如此。

他对矛盾的解答可以在打算作为《原理》第二版的草稿中找到, 此稿在他离开剑桥为谋求伦敦的一个职位时被丢弃了。我们记得他将绕日运行的有限的行星系统视为上帝为人类提供一个稳定环境的规划,虽然这种稳定是不理想的,因而上帝最后会介入,以防止引力削弱系统。恒星系统同样是稳定的;但是他争辩说,这是因为恒星在数目上是无穷的,它们的分布是(差不多是)对称的:每颗恒星最初是静止的,因为它在每个方向被其他恒星同等地拉着,所以它会继续保持如此状态。

但是只要一瞥夜天空就会看出这种对称其实是不理想的;的确,即便为了对较近恒星间表面的对称提供证据,牛顿也需要独具慧眼。但是他并不将不理想对称看作是一个问题:这是上帝的又一次有规律的干预,干预的结果是,恒星会恢复它们早先的秩序。

牛顿曾着力研究宇宙的动力学,但是使这些一样的恒星向我们发光的动力是什么呢? 1720 年左右, 这个问题由他的熟人,一个年轻的内科医生威廉·斯图克利(1687—1765)向他提出。伽利略望远镜在一个世纪前就证实银河由数不清的微小恒星并合的光所形成;但是奇怪的是,对于引起这一现象的恒星的三维分布人们几乎没有继续研究的兴趣。牛顿没有想到银河的状况反驳了他的观点,恒星宇宙其实并不是对称的。

可是,斯图克利猜测,可见的恒星会一起形成一个球状的集合体, 而银河中的恒星会在这个球体周围形成一个扁平的环——实际上，是像土星和土星环那样的恒星类似物。作为回应，牛顿暗示，他更倾向于一个无限的对称分布的恒星宇宙；对此斯图克利——不知道牛顿暗中相信的正是这个概念——反驳说，在这样的一个宇宙中，"[天空的]整个半球就会具有像银河系那样的发亮的外观"。

1721年早期，斯图克利和哈雷与牛顿共进早餐，并且讨论了天文学问题。其中必定包含了一个无限的恒星宇宙的可能性，因为几天以后，哈雷向皇家学会宣读了有关这一主题的两篇论文中的第一篇。当他的论文在《哲学会报》上发表时，牛顿的宇宙模型最后——以不具名的形式——进入了公众视野。

在一篇论文里，哈雷仔细地评述道："我所听到的另一个观点主张，如果固定的恒星数目比有限还要多，那么它们外显的球的整个外表会是发亮的。"他对斯图克利的疑虑有他自己的解决办法，但是它是有缺陷的。直到1744年，关于无限和几近对称的宇宙中光的正确分析才得以发表。瑞士天文学家德·谢塞奥（1718—1751）指出，在最近恒星的距离处，对于一定数目的恒星，有（可以这么说）空间使得任何两颗恒星不至于过分靠近；这些恒星一起填满了天球的一定（小的）面积。在两倍距离处，可容纳的恒星数目可多达之前的四倍，但是每颗恒星的亮度变成了原先的1/4，视大小也变成了原先的1/4。所以，总体来说，它们会像以前那样将天空同样的面积填满，而且具有同样的亮度水平。在三倍距离处，恒星将用光填满天空更大的面积；依次类推，直到最后整个天空群星闪耀。

人们或许会如此想（现代天文学家确实将夜空的黑暗视为提出了"奥伯斯佯谬"）。但是谢塞奥指出——如同奥伯斯在1823年所做的那样——这一推理假定所有从一颗特定恒星发出的光都到达了其目的地；即使是微小的光损失，如果在沿途的每一步都发生，很远恒星的光就会显著地减弱直至看不见为止。所以无论对于谢塞奥还是奥伯斯来说，都不存在任何佯谬。

对 19 世纪后期的天文学家来说也不存在佯谬，即使现在人们已意识到，一种拦截星光的星际介质会自我加热并开始辐射。有很多其他方法可以摆脱困难，诸如无以太真空的存在，在其中光不能通过。只有在我们的时代，夜空的黑暗才会是一个佯谬。那些给它命名的人并不知道，这一问题回溯起来，要越过奥伯斯，到谢塞奥、哈雷，并最终到内科医生斯图克利。

同时，业余观测者开始苦思银河系。1734 年，达勒姆的托马斯·赖特（1711—1786）作了一个公开讲座（或称布道），提出了他个人的宇宙论。他告诉他的听众，太阳和其他恒星围绕着宇宙的神圣中心运转。当它们这么运转时，它们占据着空间中的一层球壳，其外则是黑暗的外层空间；为了集中听众的注意，他向他们指出，他们中每一个人在死后注定要向内或向外通过这个区域。为了说服听众，他准备了直观的图片，展示了宇宙的一个截面，其中他用艺术手法描绘了地球上实际显示的太阳系和可见恒星；他说，更遥远恒星的光并在一起形成了"一个光的暗圆"——银河。

稍后，他意识到了他的错误：这样一个银河会位于通过神圣中心和太阳系的每个宇宙截面上，而实际的银河则是独一无二的。为解决这个问题，在 1750 年出版的精美插图本《一个新颖的宇宙理论》中，他大大削减了太阳和我们系统（现在他设想有许多这样的系统，每个系统有它自己的神圣中心）的其他恒星所占据的空间球壳的厚度。因此，在我们凝视空的空间以前，我们向内或向外看时，看到的只是少量近邻的（因而明亮的）恒

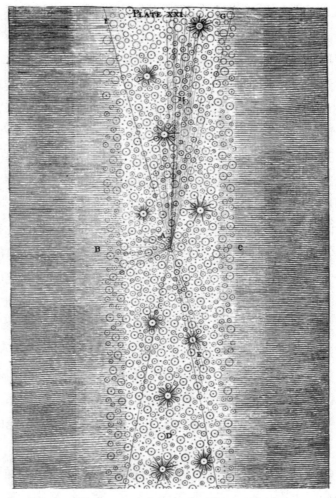

图 16  赖特所用的草图，用来帮助读者理解他所推荐的恒星系的模型。在这个想象的宇宙中，有一个被两个平行平面所限定的恒星层。在 A 的观测者，当他从层内向外看时，在 B 或 C 的方向只会看到少数近邻的（因而明亮的）恒星；但是当他沿着层面向 D 或 E 的方向看时，会看到或近或远数不清的恒星，它们的光并在一起，形成一条银河的效果。摘自托马斯·赖特的《一个新颖的宇宙理论》(1750)。

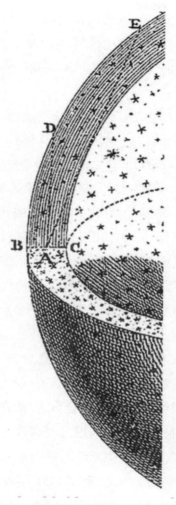

图 17 赖特推荐的太阳系所属的恒星系模型。恒星占用的空间呈球壳
形状,其半径很大,其曲率对一个位于 A 点的观测者来说难以察觉。所
以对观测者来说,可见恒星层的内、外表面近似于平行的平面。同以前
一样,向 B 或 C 的方向看时,观测者看到的只是少量近邻的因而明亮
的恒星,而当沿层面向着诸如 D 和 E 的方向看时,看到的是不可胜计
的恒星,它们的光并在一起形成一个银河的效果。

星。但是当我们沿着壳层切向看时,壳层的半径巨大,其弯曲程度难以察觉,我们看到了大量的恒星,它们的光并在一起,创造出一种乳状的效应:那就是银河的平面在我们的位置上与壳层相切。

次年,赖特的书的提要(不包括为理解他的古怪概念所需的插图)出现在了一本汉堡出的期刊上,引起了德国哲学家伊曼努尔·康德(1724—1804)的注意。康德不是没有理由地假定,必定只有一个神圣中心,它位于宇宙某个遥远的地方,而且我们的恒星系完全处于自然的秩序之中。他知道我们业已在天空中观测到乳状斑点(星云)的情况,他相信它们是其他的恒星系;但是这些系统是椭圆形的,而一个球状系统不管从哪个角度看来将永远是圆形。所以康德选择了赖特提出的另一个模型,其中围绕我们神圣中心的恒星形成了一个扁平的环。康德认为,这个环(整体上处于自然秩序之中)不应当不间断地从一边延伸到另一边,从而形成一个完整的星盘;一个星盘从侧向看来将呈椭圆形,正如业已观测到的星云一样。所以康德误以为赖特视银河为一个盘状的恒星聚合体,虽然它确实如此。

由热情研究的业余爱好者构想出的这些情形及类似的推测几乎不可能对专业的天文学家形成冲击。另一方面,他们却几乎不能忽视另一位业余爱好者在1781年发现的行星天王星。这位业余爱好者就是我们所熟知的威廉·赫歇尔,7年的战争流亡之后,这位音乐家从汉诺威来到英国。但是他的报告不够专业,而且他随口声称他是用目镜(据说其放大率甚至超过专业光学仪器)作出了这个发现,因此他成为了一个有争议的人物。

1772年，赫歇尔将他的妹妹卡罗琳从汉诺威的家庭劳作中拯救了出来，成为他做的每一件事的忠实助手。他对天文学的热心不久就左右了他们的生活。赫歇尔的雄心是了解"天的构造"。赫歇尔认识到为了观看那些遥远而暗弱的天体，他必须装备能够收集尽可能多的光的反射望远镜——换句话说，尽可能大的反射镜。他从当地铸造厂购买了圆盘，学习研磨和抛光，但他的抱负很快超出了其能力：一个3英尺的玻璃圆盘就使他放弃了。1781年，他大胆地将家里的地下室变成一个铸造车间，但是他尝试了两次都失败了，还几乎遭受了灭顶之灾。

天王星的发现使赫歇尔的赞赏者有机会代表他去游说国王，1782年赫歇尔被授予了皇家津贴，使得他能够致力于天文学研究。他搬到了温莎城堡附近，在那里，他除了满足皇室家庭和他们的客人的观天要求之外，别无其他责任。他很快建造了史上最大的望远镜之一，一架焦距长达20英尺，口径为18英寸且有一个稳定平台（这也同样重要）的反射望远镜。

有卡罗琳坐在写字桌边听觉范围以内充当抄写员，赫歇尔在此后20年中用了很多观测时间，致力于在夜空中"扫视"星云和星团。朝向南方的望远镜被安置在一定的高度，当天空在头顶缓慢地转动时，可以上下微微调整其位置使其可以扫过可能含有星云的天空中的某一条带。在他们开始时，只有约100个这样的神秘天体是已知的；到结束时，他们已经收集了2500个样品并进行了分类。

每个人都认识到，不能辨别出单颗恒星的遥远星团会呈现出星云状，如同银河那样。但是全体星云都是遥远的星团吗？其中有些会不会是由近邻的发光流体云（赫歇尔称之为"真正的星云状物质"）所形成的？如果可以看到一个星云改变形状，这

图 18 威廉·赫歇尔"巨大的"20 英尺反射望远镜(受托于 1783 年),本图摘自 1794 年发表的一张版画。他用这架仪器"扫视"天空,发现了 2500 个星云和星团。到了 1820 年,木建部分已大半朽烂,威廉的儿子约翰被迫建造了一个替代物,他将其携至好望角,将他父亲的工作拓展至了南天。

将证明它是一团近邻的云,因为一个遥远的星团太大了,不可能如此迅速地发生改变;1774 年,在其观测日志的第一页上,赫歇尔记下了猎户座大星云并不像它早先被描绘(17 世纪由惠更斯所描绘)的那样。此后几年对同一星云的偶尔观测让他确信,星云会继续变化,因而它是由真正的星云状物质构成的。但是怎样区分真正的星云状物质和遥远的星团呢?看来赫歇尔正遇到了两类星云状物质:乳状的和有斑点的。他揣测有斑点的星云状物质反映了数不清的恒星的存在。

但 1785 年他碰见的一个星云包含了单颗的恒星以及两种星云状物质;他将它解释为从观测者延伸开去的一个恒星系统。最近的恒星是单独可见的,更远的恒星看起来像有斑点的星云状物质,而最远的像乳状星云物质。因此,赫歇尔改变了他早先的立场,判定所有的星云都是星团。

但是星团意味着群集:一种吸引力或几种力——可能为牛顿引力——起着作用,将成员星愈发紧密地拉在一起。这意味着,一个星团里的恒星过去比现在更为松散;将来,它们会更紧密地抱成团。

这样一来,赫歇尔将生物学的概念引入了天文学:他像自然史学家一样收集了大量的样品并进行分类,他能够按年龄——如年轻的、中年的和老年的——进行排列。他正改变着科学的本质。

1790 年的一个晚上,他正像往常一样扫视着天空时,偶然遇见一颗恒星被一个星云状物质的晕所环绕。他认定这颗恒星定然由星云状物质凝聚而成,故而真正的星云状物质是存在的。他必须将自己恒星系统的发展理论作一回溯,使其包含一个较早的位相,在这个位相中,稀薄的散射光在引力作用之下凝聚成星云状的云,恒星就从中诞生。这些恒星形成了星团,最初是散布的,然后日益凝聚——直至最终星团自身坍塌,产生巨大的天体爆炸,爆炸产生的光开始了又一次的循环。赫歇尔的同时代人中,几乎无人有能观测到证据的仪器,所以不知道该如何应对。

威廉·赫歇尔的儿子约翰(1792—1871)将恒星天文学带入了科学的主流。当他还年轻时,他父亲说服他放弃在剑桥的事业回家,做父亲的学徒和天文学的继承人,负责重新检验和扩

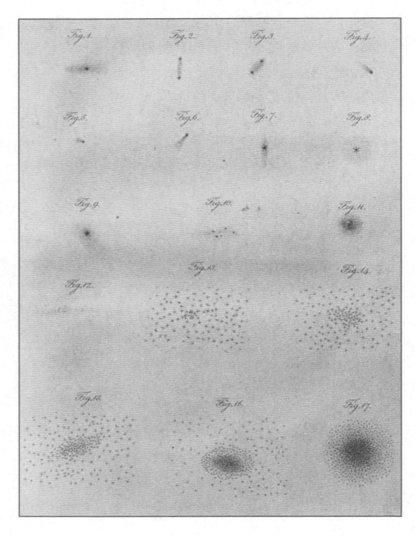

图 19 赫歇尔画的略图，表示他的星云和星团表中的天体按照成熟程度递增的序列排列：随着时间流逝，引力起的作用使星团越来越集中。摘自《哲学会刊》104卷（1814）。

大他父亲的天文样品收集。威廉的 20 英尺反射望远镜现在已因年代久远而朽烂，但他在 1822 年去世之前，监督约翰建造了一个替代物。

1825 年，约翰开始修订他父亲的在英格兰可见的星云的星表。完成之后——并且还坚定地谢绝了所有政府提供的资助——他坐船到了好望角，在那儿他花费了 4 年的时间，将他父亲的星云、双星等等的星表扩展到南天范围。他变成了（并且依然是）用一个大望远镜研究了整个天球的唯一观测者。

1838 年 3 月约翰·赫歇尔坐船返家时，作为观测者的生涯就此告终，赫歇尔家对大望远镜的垄断也是如此。那年，在爱尔兰中央地区的伯尔城堡，未来的罗斯伯爵威廉·帕森斯（1800—1867）加工和装配了部件，做成了直径 3 英尺的合成镜面。次年，他成功浇注了同样大小的单镜面，并于 1845 年完成了"帕森斯城的巨兽"。这是一架吊起在巨大石墙之间的反射望远镜，镜子口径不小于 6 英尺，每块重达 4 吨。几周之内，这架反射望远镜就揭示出，有的星云在结构上是螺旋形的。

"巨兽"被设计成一劳永逸地解决以下这个问题：是不是全部星云都是被距离伪装起来的星团？观测肉眼可见的猎户座大星云是关键，在这一点上，人们达成了共识。正好反射镜足以观测确实嵌入这个（气态的）星云中的恒星。在看到这些时，罗斯说服他自己，他正在观看一个星团并且他已胜利地将它分解成了多颗成员星。

许多人同意，这个关于最大星云的发现能够推广，并且天文学不再需要"真正的星云状物质"了。他们很快被证明是错了，但等到那时，天文学已丧失了其自主性并为从事星光分析而与物理学和化学相结合。

图 20　罗斯伯爵 6 英尺镜面的反射望远镜，受托于 1845 年。该年 4 月，罗斯用它发现了某些星云的漩涡结构。

# 后记

　　在本书中,我们追随自古以来的观测者和理论家了解了天体——它们是什么,以及它们怎样表现。不管观测者是否意识到,他们的信息都是来自从这些天体发出现在正抵达地球的光:他们观测到的是光,而不是天体本身。

　　并非所有这样的光都是一样的。例如有的恒星闪耀着灿烂的白光,而其他恒星则有微红的色彩。距离我们最近的恒星太阳的白光和颜色之间的关系是 1666 年由艾萨克·牛顿建立的。他在剑桥三一学院居室的百叶窗上开了一个小孔,太阳光的光束通过一个棱镜照了进来。不出所料,他看到了具有彩虹全部颜色的熟悉的光谱。公认的理论是,白光是简单而基本的,而颜色则源于白光的某种修改:从白光开始,对白光作了一些修改,就得到了一种颜色。通过仔细实验,牛顿发现恰恰相反,颜色才是基本的,当重新并在一起时,它们会再次形成白光。太阳光就是由彩虹的颜色组成的。

　　牛顿正研究光本身,而不是作为光源的太阳。威廉·赫歇尔是好奇地研究其他恒星光谱的第一位观测者,那架具有充分"聚光能力"的望远镜使他的工作成为了可能。早在 1783 年,当他将 20 英尺反射望远镜指向一颗亮星时,他曾几次将棱镜放在望远镜的一个或另一个目镜上,但是直到1798年4月9日,他才对6颗最亮的恒星作了短暂的研究。他发现,"天狼星的光由红、橙、黄、绿、蓝、紫红和紫色组成"。另一方面,"大角星按比例比天狼星包含更多的红色和橙色,更少的黄色"等等。但是这些差别意味着什么,他并不知道。

从对太阳光的仔细得多的分析中,答案逐渐被找了出来。1802年,威廉·海德·沃拉斯登(1766—1828)重复了牛顿的实验,用一个只有1/20英寸宽的夹缝代替了牛顿在他的百叶窗上开的小孔。他惊奇地发现,太阳光谱上有7条暗线,他将它们认定为颜色之间的分界线。但是当望远镜的制造者约瑟夫·夫琅和费在玻璃棱镜上作测试时,他惊讶地发现事实上存在着几百条这样的线。他还发现在实验室里可以产生不同的光谱,由稀疏的亮线以及它们之间的暗区所组成(与太阳和恒星的连续光谱形成对照的一种"亮线"光谱)。

在紧接着的30年间,情况逐渐明朗,其深远意义也变得显而易见。其间,化学家威廉·本生(1811—1899)和物理学家基尔霍夫(1824—1887)这两个德国人起到了核心作用。1859年他们证实了灼热的固体和液体能产生连续光谱,太阳光谱就是熟悉的例子,而灼热的气体则产生亮线光谱。[结果,当英国天文学家威廉·哈金斯(1824—1910)在1864年从天龙座的一个星云中获得了可见的亮线光谱时,他终止了延续了几个世纪的关于"真正的"(气体的)星云是否存在的辩论。]每个元素有它自己的特征线位置。令人惊奇的是,连续光谱在通过一种气体时显示为"暗线光谱",其暗线为气体的特征。结果,一旦一种元素的谱线位置在实验室中被确定以后,研究者就能够证实该元素在恒星、星云或从天体来的光所通过的任一气体中的存在或者缺失。

一个潘多拉盒子被打开了。正如伟大的美国观测者詹姆斯·基勒所评述的："光向我们显示了天体的存在也包含了它们的构造和物理状态的秘密。"1835 年,奥古斯特·康特宣告了人类知识的局限,认为我们永远无法研究天体的化学组成,这一著名的论断在情感上很难让人接受。变革是如此深刻,以至于天文学失去了它的独立性,变成了物理学(以及化学)的一个分支,正如哈金斯所说的:

> 然后,天文台第一次变成了实验室。放出有害气体的原电池在窗外排列着;一个巨大的感应线圈连同一块带有莱顿瓶的电池安放在轮子上面的架子上,以便跟随望远镜的目端位置作出调整;搁板上本生炉、真空管、化学瓶(特别是纯金属样品)靠墙排列着。

天体的性质、结构和演化变成了"天体物理学家"而不是天文学家的研究领域。开普勒给取的名称,"新天文学",再一次被援引。同时,传统天文学与天体物理学一同兴旺和发展起来。

任何一门科学的历史是永无止境的,当从业者的数目逐级上升时,学科的范围也随之扩展。19 世纪中期,天文学的变革标志着我们已开始讲述的故事的终止和(始终是)另一个故事的开始。

天文学研究现在成了科学家和工程师小组的团队工作。射电望远镜在人眼看不见的波长处截取辐射——信息。这样的望远镜组合可以等价于直径达几百公里的单个"抛物面天线"。镜面不断增大的光学望远镜建在山顶,高出绝大部分的地球大气，并且主要是在南半球，在那儿，大多数有意义的深空天体有待观测。计算机驱动了望远镜并且依靠"主动光学"对镜面的误差以及仪器上方大气中的微妙变化作出连续的补偿。新的技术极大地增加了获得的信息量。从哈勃空间望远镜及行星探测器等当代最先进的仪器中可以获取这些信息。看着从宇宙飞船通过无线电连接传输到地球的图像，就很容易理解为什么对天文学家来说这是一个最激动人心的时刻。

# 索引

（条目后的数字为原文页码）

**104**

## Q

## R

## S

**109**

Michael Hoskin

# THE HISTORY OF ASTRONOMY

## A Very Short Introduction

# Contents

# List of illustrations

The publisher and the author apologize for any errors or omissions in the above list. If contacted they will be pleased to rectify these at the earliest opportunity.

# Chapter 1
# The sky in prehistory

Historians of astronomy work mainly with the surviving documents from the past (fragmentary in quantity from antiquity, overwhelmingly bulky from recent times), and with artefacts such as instruments and observatory buildings. But can we discover something of the role the sky played in the 'cosmovision' of those who lived in Europe and the Middle East *before* the invention of writing? Could there even have been a prehistoric science of astronomy, perhaps one that enabled an elite to predict eclipses?

To answer these questions we rely primarily on the surviving stone monuments – their alignments, their relationships to the landscape, and the (usually ambiguous) carvings we find on some of them. The underlying problem of methodology is at its most acute when we are dealing with a monument that is unique. Stonehenge, for example, faces midsummer sunrise in one direction and midwinter sunset in the other. How can we be sure that an alignment that to us is of astronomical significance was chosen by Stonehenge's architects for this very reason? Did it have some quite different motivation, or even occur purely by chance? To take another example, a monument built around 3000 BC that faces east may have been oriented on the rising of the Pleiades, a bright cluster of stars in the constellation Taurus. It may have faced midway between midsummer and midwinter sunrise. Perhaps there was a sacred mountain in that direction. Or the orientation may have been chosen simply to take advantage of the slope of the ground. How can we decide which of these, if any, was in the minds of the builders?

We are on safer ground when we are dealing with a large number of monuments spread over a wide area. Archaeologists of western Europe study the communal tombs of the late Stone Age (the Neolithic), by which time the nomadic life of the hunter-gatherer had been replaced by the more settled existence of the farmer. Such tombs were to serve the needs of the clan for many years, and so they had an entrance through which additional bodies could be introduced as need arose. We can define the orientation of the tomb to be the line of sight of the bodies within as they 'look' out through the entrance.

In central Portugal there are many such tombs with a very characteristic and instantly recognizable shape, structures built by people with shared customs. They are scattered over a mountainless region that measures some 200 km from east to west and a similar distance from north to south; yet every single one of the 177 tombs that the writer has measured faces easterly, within the range of sunrise.

Not only that, but the directions in which the Sun rose in the autumn and winter predominate. Now we know from written records that Christian churches in many countries have traditionally been oriented to face sunrise (twice in the year), because the rising Sun is a symbol of Christ; and the builders often ensured this would happen by laying the church out to face sunrise on the very day that construction began. Suppose the Neolithic builders of these tombs followed a similar custom; suppose they too saw the rising Sun as a symbol of a life to come. Then, since it was doubtless in the autumn and winter, after the harvest, that they were free to dedicate themselves to such work, we would expect to find just such a pattern of orientations as we do in fact find; and it is difficult to imagine any other way of accounting for the striking

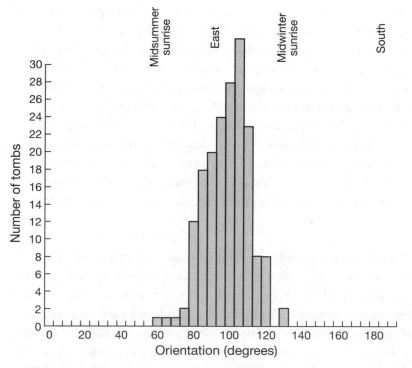

**1. A histogram showing the orientations of 177 seven-stone-chambered tombs of central Portugal and adjacent regions of Spain. When the horizon altitudes have been taken into account, we find that every single tomb faced sunrise at some time of the year, most of them in the autumn months, when we might expect the constructors to have been at liberty to undertake such work. This is consistent with a custom of orienting tombs to face sunrise on the day construction began, as was later practised with Christian churches in England and elsewhere.**

pattern of orientation. It seems reasonable, therefore, to infer that the Neolithic builders oriented their tombs on sunrise.

If this is so, then we have evidence that the sky played a role in the Neolithic cosmovision, just as it played (and plays) a role in the cosmovision of church builders. But this has nothing to do with 'science'. Claims that there was indeed an authentic science of astronomy in prehistoric Europe were made a generation ago by Alexander Thom, a retired engineer who took it upon himself to survey hundreds of stone circles in Britain. According to Thom, the prehistoric builders located the circles so that from them, the Sun (or Moon) could be seen rising (or setting) behind a distant mountain on a significant day – the winter solstice, for example, in the case of the Sun. For several days around the solstice, the Sun rises (or sets) at almost the same position on the horizon, and only with a very accurate instrument can the actual day of the solstice be identified. According to Thom, the prehistoric elite used circles in combination with distant mountains to form instruments many miles in extent; and, with the knowledge of solar and lunar cycles that this gave them, they could predict eclipses, and thus confirm their ascendancy over those around them.

Thom's work aroused enormous interest and, of course, controversy. However, a reinvestigation of his sites concluded that he had singled out those distant mountains that he knew would fit his ideas, and that such alignments could well have occurred by chance, and have been of no concern whatever to the prehistoric builders. Few now give credence to Thom's speculations, although anyone attempting to understand the cosmovision of a prehistoric people owes him a debt of gratitude for drawing attention to such issues.

We can be sure that in prehistoric times the sky served the practical needs of at least two groups: navigators and farmers. Today, in the Pacific and elsewhere, navigators use the Sun and stars to find their way, and no doubt prehistoric sailors in the Mediterranean did the same; but few if any traces of this survive. Concerning the agricultural calendar – farmers have always needed to know when to plant and when to harvest – we have some clues. Even today there are places in Europe where farmers avail themselves of celestial signals of the type described for us in *Works and Days* by the Greek poet Hesiod (*c.* 8th century BC). Each year the Sun completes a circuit of the stars, and there is therefore a period of some weeks when a given star – Sirius, for example – is too close to the Sun to be visible in the daytime. But the Sun moves on, and the day comes when Sirius can be glimpsed in the dawn sky: its 'heliacal rising'. Hesiod describes the sequence of heliacal events used by the farmers of his day for their calendar, and this must encapsulate the knowledge and experience assembled over the preceding centuries. Surprisingly, there seems to be much earlier evidence of just such a sequence inscribed on pillars of the temple of Mnajdra on Malta, which dates from around 3000 BC. My colleagues and I found rows of incised holes that seem to be tallies, and on analysing the counts we found that they may well indicate the numbers of days between one important heliacal rising and the next. As we shall see, the heliacal rising of Sirius was soon to play a pivotal role in the calendar of nearby Egypt.

# Chapter 2
# **Astronomy in antiquity**

The origins of modern astronomy first emerged from the mists of prehistory in the 3rd and 2nd millennia before Christ, in the increasingly complex cultures that developed in Egypt and Babylon. In Egypt the effective administration of a far-flung kingdom depended upon a well-established calendar, while rituals called for the ability to tell the time at night, and for the capacity to orientate monuments – pyramids – in the cardinal directions. In Babylon the security of the throne, and therefore of the state, depended upon the correct reading of omens, including those seen in the sky.

Calendars were, and are, awkward to formulate because there is no exact number of days in either the lunar month or the solar year, and likewise no exact number of months in the year; our own extraordinary jumble of month-lengths is a symptom of the problems nature poses for the calendar maker. In Egypt life was dominated by the annual flooding of the Nile, and a solution to the calendar problem was found when it was noticed that this flooding took place around the day when Sirius rose heliacally – when the star appeared in the dawn sky after an absence of weeks. The star's rising could therefore be used to anchor the calendar.

Each year consists of 12 lunar months and about 11 days, and the Egyptians devised a calendar in which Sirius would *always* rise in the 12th month. If in any given year the star rose early in the

12th month, well and good: it would rise next year in the same month. But if it rose late in the 12th month, then unless action was taken, the following year it would rise after the month had ended. To prevent this happening, an extra, or 'intercalary', month would be declared for the current year.

Such a calendar was satisfactory for religious festivals, but not for the administration of a complex and highly organized society, and so for civil purposes a second calendar was devised. It was ruthlessly simple: every year consisted of exactly 12 months, each of three 'weeks' of ten days, together with an extra five days at the end of the year to bring the total number of days to 365. As the seasonal year is in fact a few hours longer (which is why we have leap years), this civil calendar slowly cycled through the seasons; but this was considered an acceptable price to pay for the administrative convenience of an unchanging pattern.

Since there were 36 'weeks' of ten days, 36 star groups or 'decans' were selected around the sky so that a new decan rose heliacally every ten days or so. As dusk fell on any given night a number of decans would be visible overhead, and during the night new ones would appear on the horizon at regular intervals, so marking the passage of time.

The sky played a profound role in Egyptian religion, for deities were present there in the form of constellations, and immense labour was expended on Earth to ensure that the reigning pharaoh would one day join them. We see one aspect of this in the almost precise north–south alignments of the funerary pyramids of pharaohs of the 3rd millennium, and there has been much debate as to how this was achieved. A clue comes from the (tiny) errors in the alignments, for these errors change systematically with the dates of construction. It has recently been suggested that the Egyptians may have referred to an imaginary line joining two particular stars that were seen above the horizon at all times (circumpolar stars), and have taken north to be the direction towards this line at the exact moment when the line was vertical. If so, the slow movement of the celestial north pole due to the wobble of the Earth's axis (called precession) would account for the systematic errors.

The Egyptians were handicapped by the primitive condition of their geometry and arithmetic, and this precluded them from developing an understanding of the more subtle movements of the stars and planets. In particular, their arithmetic operated almost exclusively with fractions that had the number one in the numerator.

By contrast, 2,000 years before Christ the Babylonians developed a brilliant technique for arithmetical notation, and this was the basis of their remarkable achievement in astronomy. A scribe would take a soft clay tablet the size of a man's hand, and impress his stylus on it edgeways to denote a 1, and flatways to denote 10. Doing this as often as necessary, he would write numbers from 1 to 59; but for 60 he would use the symbol for 1, much as we do in writing 10, and similarly for $60 \times 60$, $60 \times 60 \times 60$, and so on. There were no limits to the accuracy and versatility of the numbers that could be written in this sexagesimal system of notation, and even today we continue to write angles in sexagesimal degrees, and to reckon time in hours, minutes, and seconds.

Babylonian court officials were on the alert for omens of all kinds – the entrails of sheep were of special interest – and they kept records of any unwelcome events that ensued, so as to learn from experience: when the omen occurred again in the future, they would know the nature of the impending disaster of which the omen was a warning, and so the appropriate ritual could be performed. This led to the compilation of a vast compendium of 7,000 omens that had taken definitive form by 900 BC.

Soon thereafter the scribes began systematically to record astronomical (and meteorological) phenomena, in order to refine their prognostications. For seven centuries this continued, and gradually cycles in the movements of the Sun, Moon, and planets

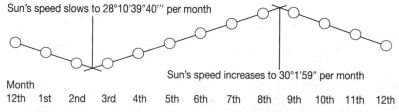

Sun's speed slows to 28°10'39"40''' per month

Sun's speed increases to 30°1'59" per month

Month
12th 1st 2nd 3rd 4th 5th 6th 7th 8th 9th 10th 11th 12th

**2. A repesentation in modern terms, with the values found in a tablet for 133/132 BC, of the second Babylonian approximation of the speed of the Sun against the background stars. In this artificial but arithmetically convenient formulation, the speed is imagined to increase by the same amount each month for six months, and then to decrease similarly for the next six months. This was found to yield acceptably accurate results.**

began to emerge from the records. With the help of their sexagesimal notation, the scribes devised arithmetical techniques for using these cycles to predict the future positions of the celestial bodies. For example, the Sun's movement against the background stars accelerates for one half of the year, and decelerates for the other half. The Babylonians devised two techniques for approximating to this movement: either they assumed one uniform speed for half the year and another uniform speed for the remaining half; or they assumed a steady increase in speed for half the year and a steady decrease for the remaining half. Both were no more than artificial approximations to reality, but they did the job.

Of Greek astronomy prior to the 4th century BC our knowledge is fragmentary in the extreme, for few writings survive from this period, and much of what we have is in the form of citations by Aristotle (384–322 BC) of opinions he is about to attack. Two aspects, however, stand out: first, the emergence of an attempt to understand nature in purely natural terms, without recourse to the supernatural; and second, the recognition that the Earth is a sphere. Aristotle rightly points out that the shadow of the Earth cast on the Moon during an eclipse is invariably circular, and that only if the Earth is a sphere can this be the case.

Not only did the Greeks know the shape of the Earth, but

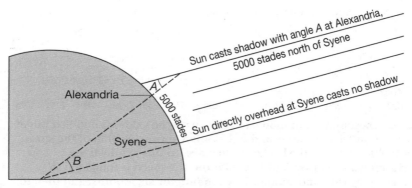

Alexandria

5000 stades

Syene

A

B

Sun casts shadow with angle A at Alexandria,
5000 stades north of Syene

Sun directly overhead at Syene casts no shadow

**3. The geometry used by Eratosthenes to measure the Earth. Angles *A* and *B* are equal.**

## Eratosthenes's measurement of the circumference of the spherical Earth

Eratosthenes believed that at what is now Aswan, the Sun was overhead at noon on midsummer's day, whereas at Alexandria, thought to be 5,000 stades due north of Aswan, it was one-fiftieth of a circle from being directly overhead. This being so, simple geometry showed that the circumference of the Earth was 50 times 5,000 stades. The modern equivalent of the stade is debated, but there is no doubt that the value of 250,000 stades was approximately correct.

Eratosthenes (*c*.276–*c*.195 BC) arrived at an excellent estimate of the Earth's actual size. Ever since then, everyone with a modicum of education has known that the Earth is spherical.

So, it seemed, was the sky. Furthermore, we always see exactly half the celestial sphere, and therefore the Earth must be at its very centre. And so developed the classic Greek model of the universe: a spherical Earth at the centre of a spherical cosmos.

Aristotle, in voluminous writings that were still being taught in

Cambridge in Isaac Newton's day, contrasted the terrestrial region at the centre of the cosmos – extending almost as far as the Moon – with the celestial region that lay beyond. In the terrestrial region there was change, life and death, coming to be and passing away. At the very centre was the sphere of Earth; around it, the shell of water, then the shell of air, and finally the shell of fire. Bodies were made of these elements in varying proportions. Left to itself, a body would move in a straight line, either towards the centre or away from it, in order to reach the distance from the centre appropriate to its elemental makeup: thus stones, being primarily earthy, fell down towards the centre, whereas flames rose towards the sphere of fire.

Immediately beyond the sphere of fire was the beginning of the celestial region, where the movements were cyclic (never rectilinear), and therefore there was no true change. Highest in the sky was the rotating sphere of the innumerable 'fixed' stars, so-called because they never altered their positions relative to each other. The stars that were not fixed numbered just seven: the Moon (clearly the nearest of all), the Sun, Mercury, Venus, Mars, Jupiter, and Saturn. These moved against the background of the fixed stars, and because their movements were forever changing – indeed, the five lesser bodies actually reversed direction from time to time – they were known as 'wanderers', or 'planets'. Aristotle's teacher Plato (427–348/7 BC), a mathematician by instinct, had seen the planets as possible disproof of his belief that we live in a cosmos governed by law: but could it perhaps be shown that the planets were in fact as regular in their movements as the fixed stars, the sole difference being that the laws governing the planetary movements were more complex, and not immediately obvious?

The challenge was taken up by the geometer Eudoxus (c.400–c.347 BC), who formulated for each planet a nest of either three or four concentric spheres, which he used in a mathematical demonstration that the planets' movements were lawlike after all. Each planet was imagined as being on the equator of the innermost sphere, which rotated with uniform speed, carrying the planet with

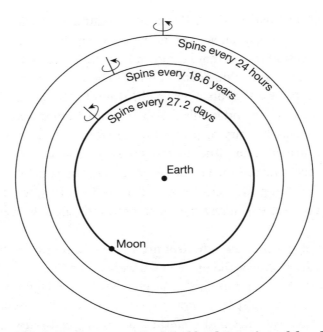

Spins every 24 hours

Spins every 18.6 years

Spins every 27.2 days

Earth

Moon

**4. The mathematical patterns displayed by the motion of the planet Moon, according to Eudoxus. The Moon is imagined to be on the equator of the innermost sphere, which rotates once a month. The poles of this sphere are embedded in the next sphere, which rotates every 18.6 years, a period familiar from eclipse cycles; and the poles of this sphere are embedded in the outermost sphere, which rotates daily.**

it. The poles of this sphere were thought of as embedded in the next sphere and carried round by it as it too uniformly rotated, and so on for the third and (in the case of the lesser planets) the fourth. The angle of the axis of each sphere was carefully chosen, as was its speed of rotation, with the outermost sphere in each case generating the daily path of the planet around the Earth. The spheres of the Moon, for example, rotated with uniform speed every 24 hours, every 18.6 years, and every 27.2 days, respectively, and so the resultant motion of the Moon reflected all three periods.

For each of the five lesser planets, two of the spheres rotated with equal and opposite speeds about axes that differed only slightly, and these spheres by themselves would give the planet a motion in a figure-of-eight; this allowed the complete nests of four spheres to generate backward motions from time to time.

So far, so good. But in these geometric models the backward (retrograde) motions of the lesser planets repeated themselves with complete regularity, and clearly did not reproduce the erratic movements of the planets that we actually observe in the sky. Furthermore, the models forced each planet to remain at a constant distance from the central Earth, whereas in the real world the lesser planets vary considerably in brightness and so presumably in distance from us. Such shortcomings would have been anathema to a Babylonian, but the models were sufficiently promising to satisfy Plato's generation that the cosmos was indeed lawlike, even if its laws had yet to be elucidated completely.

Aristotle was exercised about a quite different limitation: the spheres of the models were constructions in the minds of mathematicians, and did not explain in physical terms how the planets come to move as we observe them to do. His solution was to convert the mathematical spheres into physical reality, and to combine them to make one composite nest for the entire system. The daily rotation of the outermost sphere of all, that of the fixed stars, now sufficed to impose a daily rotation on every planet within, so the outermost sphere of each planetary nest could be discarded. However, the spheres special to an individual planet would transmit their motion down through the system, unless steps were taken to prevent this; and Aristotle therefore interpolated additional spheres with opposite motions in the appropriate places, to cancel out any unwanted rotations.

The resulting Aristotelian cosmology – a central terrestrial or sublunary region where there was coming to be and passing away, and an outer celestial region whose spheres generated the cyclic movements of the fixed stars and planets – was to dominate Greek, Islamic, and Latin thought for the better part of two millennia.

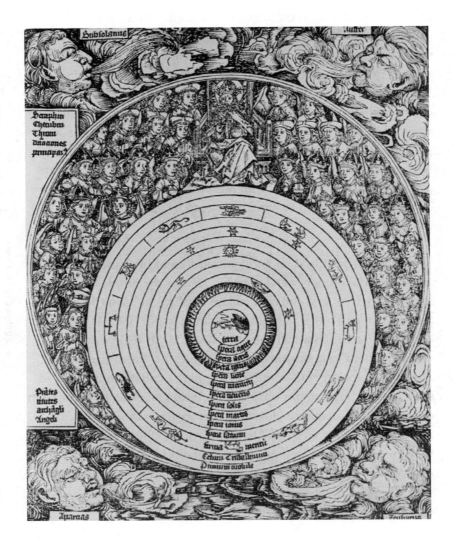

**5.** Aristotle's cosmos in Christian guise, as portrayed in the *Nuremberg Chronicle* of 1493. At the centre are the four elements (earth, water, air, and fire); then come the spheres of the planets (Moon, Mercury, Venus, Sun, Mars, Jupiter, and Saturn), followed by the spheres of the firmament of stars, the crystalline heaven, and the First Mover. Outside we see God enthroned with the nine orders of angels.

Yet its inflexibility and the resulting gap between theory and observation gained significance almost immediately, as Aristotle's pupil, Alexander the Great, conquered much of the known world and Greek geometrical astronomy began to merge with the arithmetical and observation-based astronomy of the Babylonians. Uniform circular motions continued to be seen by Greek astronomers as the key to understanding the universe, but now they were to be employed with more flexibility and with greater concern for observational fact.

Around 200 BC the geometer Apollonius of Perga developed two geometrical tools that supplied this flexibility. In one, the planet moved uniformly on a circle, but the circle was now *eccentric* to the Earth.

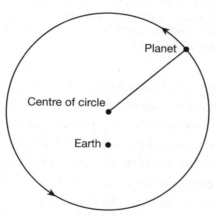

**6. In an eccentric circle, the planet moved as usual in a circle around the Earth with uniform speed; but because the Earth was not at the centre of the circle, the planet's speed appeared to vary.**

As a result, the planet would appear to move faster when its path brought it nearer the Earth, and slower when it was away on the far side of its orbit. In the other, the planet was located on a little circle, or *epicycle*, whose centre was carried around the Earth on a *deferent* circle.

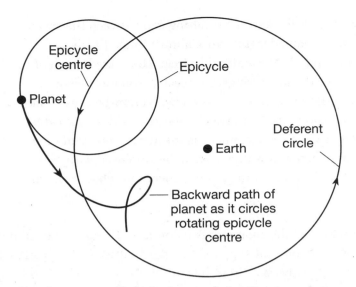

**7. An epicycle was a small circle that carried a planet moving around the circle with uniform speed. The centre of the circle likewise moved with uniform speed on a deferent circle around the Earth. The diagram indicates how the apparent backwards motion observed from time to time in the five lesser planets could be imitated in the model if the two speeds were appropriately chosen.**

It is easy for us to appreciate the value of this device, because – as seen by us – Venus (for example) orbits the Sun, which in turn orbits the Earth. Astronomy, one might say, was on the right track; not only that, but on a most promising one, for repeated refinement of the various quantities (parameters) involved would lead to encouraging progress, but never to total success – until at last Kepler abandoned circles in favour of ellipses.

The first to employ these devices was Hipparchus, who made observations at Rhodes between 141 and 127 BC. Although all but one of his works are lost, having become obsolete when later subsumed into the *Almagest* of Ptolemy, we are informed about his achievements by what we read of him in the *Almagest*. It was through Hipparchus that the geometrical astronomy of the Greeks began to incorporate precise parameters derived from the long centuries during which the Babylonians had kept their observational records. Hipparchus compiled a list of lunar eclipses observed at Babylon from the 8th century BC, and these records were crucial to his study of the motions of the Sun and Moon, for it is during an eclipse that these two bodies are exactly in line with the Earth. Hipparchus adopted the Babylonian sexagesimal system for writing numbers, and divided the ecliptic and other circles into 360 degrees. He succeeded in reproducing the solar motion by means of a single eccentric circle, and Ptolemy was to take over this model almost unchanged. He was less successful with the Moon, while the lesser planets he left to his successors.

Hipparchus's single most important discovery was that of the precession of the equinoxes, the slow movement among the stars of the two opposite places where the Sun crosses the celestial equator. The spring equinoctial point is used by astronomers to define their frame of reference, and the movement of this point implies that the measured position of a star varies with the date of measurement.

Hipparchus also compiled a star catalogue, but this is lost; the only surviving catalogue from antiquity is the one in the *Almagest*. Whole forests have been sacrificed to the debate among historians as to whether Ptolemy himself observed the positions given in his catalogue, or whether he took the positions as observed by Hipparchus and simply converted them to his own epoch by correcting for precession.

The three centuries that separated Hipparchus and Ptolemy were a dark age for astronomy; at least, Ptolemy seems to have despised whatever was done in that time and tells us little about it. Most of what information we have has been recovered from later writings in Sanskrit, for Indian astronomy was very conservative and its writers preserved what they had learned from the Greeks. But when we come to the *Almagest* itself, we are on safer ground. Of the author's life we know little, but he reports observations he made between AD 127 and 141 in the great cultural centre of Alexandria, and so he

17

cannot have been born much later than the beginning of the 2nd century. He may well have spent his adult life in Alexandria, home of the great museum and library, and he is (like Hipparchus) an example of a Greek astronomer who flourished in a place remote from the Greek mainland but close to the irreplaceable observational records of the Babylonians.

The *Almagest* is a magisterial work that provided geometrical models and related tables by which the movements of the Sun, Moon, and the five lesser planets could be calculated for the indefinite future. Written half a millennium after Aristotle, when Greek civilization had almost run its course, it synthesizes the Graeco-Babylonian achievement in mastering the movements of the wanderers. Its catalogue contains over 1,000 fixed stars arranged in 48 constellations, giving the longitude, latitude, and apparent brightness of each. As the works of earlier authors, notably Hipparchus, were now obsolete, they vanished from the face of the Earth, and the *Almagest* would dominate astronomy like a colossus for 14 centuries to come.

But there were problems ahead. Aristotelian cosmology had explained the heavens in terms of spheres concentric with the Earth, and philosophers felt comfortable with such spheres and their uniform rotations. However, Apollonius and Hipparchus had introduced eccentres and epicycles that violated this convention; true, the planets in such models moved uniformly on their circles, but not about the Earth. This was bad enough, but Ptolemy found it necessary to use a still more questionable device in order to 'save the appearances' of the planetary motions with economy and accuracy: the *equant point*.

In a planetary model, the equant was the point symmetrically opposite the eccentric Earth, and the planet was required to move on its circle so that from the equant point it would *appear* to be moving uniformly across the sky. But since the equant point was *not* at the centre of the circle, to do this the planet would have to vary its

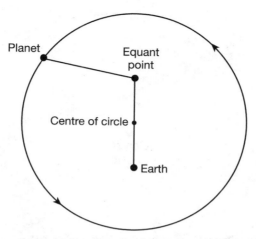

**8. The equant point was the mirror image of the (eccentric) Earth, and the planet was supposed to move in such a way that from the equant point it would appear to be moving uniformly. In fact, therefore, the planet moved non-uniformly.**

speed. Ptolemy was an astrologer anxious to know the positions of the planets at all times (his *Tetrabiblos* is a classic of astrology); and accurate prediction – however questionable the means used to achieve it – had higher priority than fidelity to the philosophical norm that all motions on circles be uniform. For him, as for the Babylonians, accuracy, not truth, was the primary consideration.

Kepler's laws of planetary motion reveal to us just why the equant was such a useful geometrical tool.

In conformity with the first two laws, the Earth (for example) moves in its orbit around the Sun in an ellipse, with the Sun at one of the two foci; and the line from the Sun to the Earth traces out equal areas in equal times. Therefore, when the Earth in its orbit is near the Sun it moves faster than usual, and when it is far from the Sun (and therefore near the other, 'empty' focus of the ellipse) it moves more slowly. Viewed from the empty focus, the Earth's speed across the sky will appear almost uniform: for when near the Sun and far from the empty focus the Earth is moving faster than usual, and this is masked by its greater distance from the empty focus; when near

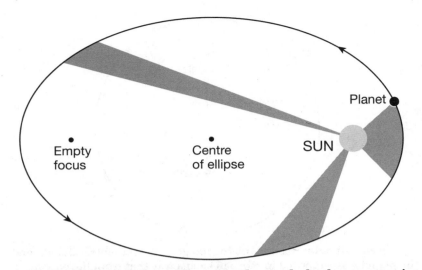

9. Kepler's first two laws enable us to understand why the equant point was a useful tool. They imply that a planet orbiting the Sun in an ellipse moves faster when near the Sun and slower when near the other focus of the ellipse. As a result, when viewed from this 'empty' focus, the planet's movement will appear approximately uniform. In this diagram the ellipticity of the orbit has been greatly exaggerated.

the empty focus the Earth is moving more slowly than usual, and this is masked by its nearness to the empty focus. In other words, Kepler teaches us that to an approximation, the Earth's speed across the sky as viewed from the empty focus is indeed almost uniform; and the empty focus is the counterpart in the Keplerian ellipse of the equant in the Ptolemaic circle.

In the universities of the later Middle Ages, students would be taught Aristotle in philosophy and a simplified Ptolemy in astronomy. From Aristotle they would learn the basic truth that the heavens rotate uniformly about the central Earth. From the simplified Ptolemy they would learn of epicycles and eccentrics that violated this basic truth by generating orbits whose centre was not the Earth; and those expert enough to penetrate deeper into the Ptolemaic models would encounter equant theories that violated the (yet more basic) truth that heavenly motion is uniform. Copernicus, among many others, would be shocked.

Nevertheless, with the models of the *Almagest* – whose parameters would be refined over the centuries to come – the astronomer, and the astrologer, could compute the future positions of the planets with economy and reasonable accuracy. There were anomalies – the Moon, for example, would vary its apparent size dramatically in the Ptolemaic model but does not do so in reality, and Venus and Mercury were kept close to the Sun in the sky by a crude *ad hoc* device – but as a geometrical compendium of how to grind out planetary tables, the *Almagest* worked, and that was what mattered.

In his *Planetary Hypotheses*, composed after the *Almagest*, Ptolemy presented his cosmology. Like earlier Greek cosmologists, he thought it reasonable to assume that the longer a planet takes to orbit the heavens against the background of the stars – that is, the less difference there is between its motion and the regular daily motion of the stars – the nearer it is to the stars. This being so, Saturn, with its period of 30 years, was nearest to the fixed stars and furthest from Earth, followed by Jupiter (12 years) and Mars (two years) in descending order. The Moon (one month) was nearest to Earth. But what of the Sun, Venus, and Mercury, which keep company as they move among the stars and therefore all have the same period of one year? Because of the Sun's dominance in the sky, and because some planets kept it company while others did not, the Sun was traditionally taken to be the middle one of the seven planets, located immediately below Mars and separating the planets that kept it company from those that did not. The order of Venus and Mercury had long been disputed; Ptolemy placed Mercury below Venus, on little more than the toss of a coin.

Having established the order of the planets – by reasoning that varied from the plausible to pure guesswork – Ptolemy now made the assumption that every possible height above the Earth was from time to time occupied by one particular planet, and only one. Thus the greatest height of the Moon (Ptolemy had an argument to show that this was 64 Earth radii) equalled the least height of the next planet, Mercury. The geometrical model for Mercury specified the ratio between its least and greatest heights, and multiplying 64 Earth radii by this ratio gave the greatest height of Mercury; and so on. That is, the geometrical models gave the ratios between the least and greatest heights of each successive planet, and the maximum lunar height of 64 Earth radii calibrated the entire system. The fixed stars, which lay at the maximum height of Saturn, were 19,865 Earth radii above us, or some 75 million miles: the Ptolemaic universe was impressively large.

Hipparchus had begun the process of employing the tools provided by the Babylonian tradition – arithmetical versatility and the use of long centuries of observation to determine parameters of exceptional accuracy – to pursue the central ambition of Greek geometrical astronomy: to reproduce the entire orbit of each planet by means of a geometrical model based on the fundamental cosmological principle of uniform circular motion. Ptolemy had carried the process to fruition, albeit with some compromises. The models of the *Almagest* would be refined time and again in the years to come; but only after 14 centuries, and the invention of printing, would a mathematical astronomer of equal competence consider its defects so fundamental as to call for a reformation.

# Chapter 3
# Astronomy in the Middle Ages

It was in AD 622 that the prophet Mohammad fled Mecca to Medina, and before long the new religion of Islam had spread across the whole of North Africa and into Spain. Islam made specific demands upon the skills of astronomers. The month began with the new moon – not when Sun, Moon, and Earth were geometrically aligned, but two or three days later, when the crescent was seen by human eyes. Could this be regularized, so that neighbouring villages would agree on the beginning of the new month, even when the sky was clouded? The hours of prayer were set by the altitude of the Sun as it traversed the sky, and the need to determine these hours correctly eventually led to the institution of the office of *muwaqqit*, or mosque timekeeper, so giving astronomers a secure and respected position in the community. And the determination of the local direction of Mecca, the *qibla*, which dictated the layout of mosques and graveyards and much else, posed a challenging problem that *muwaqqits* and other astronomers sought to solve.

Long before the arrival of Islam, the great centre of learning in Alexandria had fallen on troubled times. The *Almagest* itself was to find its way to Constantinople, and in the 9th century a copy was purchased by emissaries from Baghdad, where the youthful and vibrant Muslim culture had woken up to the intellectual treasures surviving in the Greek language. At

Baghdad it was translated by a team working in the House of Wisdom, first from Greek into Syriac and then from Syriac into Arabic. Other copies in Constantinople would gather dust, unread, until in the 12th century the emperor presented one as a ceremonial gift to the King of Sicily, where it was translated into Latin.

Despite the censure in the Koran, astrology flourished in the Muslim world at every level of society; and those astrologers who were more than mere fortune tellers based their predictions on tables of planetary positions. The success of the models of the *Almagest* was undisputed, but these models incorporated parameters that could be determined with ever greater accuracy as the centuries passed – and Ptolemy himself had explained how to do this. At first the measuring instruments used by astronomers for this purpose were modest in size, but as the observers' ambitions grew so did the size of their instruments, and they looked to patrons to pay for their construction and housing.

At times, however, this aroused the hostility of the religious authorities, and a patron's death – or even his loss of nerve – could bring astronomical observation to an end. In Cairo the construction of an observatory began in 1120 on the order of the vizier, but in 1125 his successor was killed by command of the caliph, his crimes included communication with Saturn, and the observatory was demolished. In Istanbul an observatory for the astronomer Taqi al-Din was completed by Sultan Murad III in 1577 – as it happened, just in time for observations of a bright comet. Taqi al-Din, doubtless with an eye to his own prosperity, interpreted the apparition as boding well for the sultan in his fight against the Persians. But events turned out otherwise, and in 1580 religious leaders convinced the sultan that it was inviting misfortune to pry into the secrets of nature. The sultan therefore ordered the observatory to be destroyed 'from its apogee to its perigee'.

Only two Islamic observatories enjoyed more than a brief existence. At Maragha, the present-day Maragheh in northern Iran, construction of an observatory for the distinguished Persian astronomer Nasir al-Din al-Tusi (1201–74) was begun in 1259 by Hulagu, the Mongol ruler of Persia. Its instruments included a 14-foot-radius mural quadrant (an instrument for measuring altitudes, attached to a wall aligned in the north–south direction) and an armillary sphere (used for other measurements of position) with circles five feet in radius. With the help of these instruments, a team of astronomers completed in 1271 a *zij*, or collection of astronomical tables with instructions for their use, in the tradition of Ptolemy's own *Handy Tables*. But in 1274 al-Tusi left Maragha for Baghdad, and although observations at the observatory continued into the next century, its creative period was already over.

The other major observatory of Islam enjoyed the advantage that the prince himself was an enthusiastic member of staff. At Samarkand in central Asia, Ulugh Beg (1394–1449), a provincial governor who was to succeed to the throne in 1447, began construction of a three-storey building in 1420. Its chief instrument, built on the principle that 'bigger is better', was a form of sextant no less than 130 feet in radius. This was mounted out of doors between marble walls aligned north–south, and the range of the instrument was chosen so that it could be used to observe the transit of the Sun, Moon, and the other five planets. The great achievement of Samarkand Observatory was a set of astronomical tables that included a catalogue of over 1,000 stars. Much earlier, the Baghdad astronomer Abd al-Rahman al-Sufi (903–86) had prepared a revision of Ptolemy's star catalogue in which he gave improved magnitudes and Arabic versions of the identifications; but he had left the stars themselves and their often inaccurate relative positions unchanged, and so Ulugh Beg's was to be the single important star catalogue of the Middle Ages. Samarkand Observatory fell into disuse soon after the murder of Ulugh Beg in 1449.

Observatories were for an elite, but every astrologer needed to make observations, and this became possible through the development of the astrolabe, an ingenious portable computer and observing instrument that had its roots in antiquity. The typical astrolabe consisted of a brass disc that could be suspended by use of a ring at the top edge. One side of the astrolabe was for observations of the angular altitude above the horizontal of a star or planet; the observer suspended the instrument and looked at the heavenly body along a sighting bar, and then read the angle on a scale around the circumference. The other side of the disc represented the celestial sphere projected on to the plane of the equator from the south celestial pole.

10. A 14th-century astrolabe preserved at Merton College, Oxford

Each line from the pole intersected the celestial sphere in one further point, and it intersected the plane of the equator (in a single point); the latter point was the 'projection' of the former. Because the brass disc was of course of finite size, and because the heavens south of the Tropic of Capricorn were of no practical interest, the projected skies extended from the north celestial pole (represented by the centre of the disc) as far as this Tropic but not beyond.

Circles of equal altitude at the latitude of the observer projected into cirles that were engraved on the disc, along with much else. So far, so good; but the stars of the rotating heavens also needed representation. This was achieved by means of a brass sheet that bore indications of the locations of the principal stars but was otherwise cut away as much as possible to reveal the coordinate circles below. This sheet rotated about the central point of the disc underneath, just as the stars rotate about the north celestial pole. The sheet also contained a representation of the ecliptic path of the Sun, and the observer needed to know (and mark) the Sun's current position on it.

This done, a single observation – typically, of the altitude of a star at night or of the Sun by day – would allow the observer to rotate the sheet into its correct current position, by moving it until the star (or the Sun) lay over the appropriate coordinate circle. This done, the entire heavenly sphere was now in position, and many questions could be answered – for example, which stars were currently above the horizon and what was the altitude of each. Time could be determined by aligning the Sun with a scale on the perimeter of the disc and reading off the hour on the scale. This was possible irrespective of whether it was by observation of the Sun or of a star that the sheet had been positioned: the astrolabe was a clock that could be used to tell the time night and day throughout the 24 hours.

A wide range of other information could be obtained easily from the astrolabe. For example, to determine the hour at which a given star would rise, the astronomer would rotate the sheet until the star was above the circle of zero altitude, and then read off the time. The astrolabe was a simple, ingenious, and versatile device that encouraged quantitative observation of the heavens.

A *zij* had been composed as early as the first half of the 9th century, in the House of Wisdom at Baghdad, by al-Khwarizmi, a corruption of whose name gives us the word 'algorithm'. It made use of the parameters and computational procedures contained in a Sanskrit astronomical work that had been brought there around 770. In a later version, the *zij* was to be translated into Latin in the 12th century, and so became a vehicle by which Indian astronomical methods reached the West. By making possible the prediction of future planetary positions, *zijes* supplied the needs of the practising astronomer/astrologer, and great numbers of these tables were produced, often using parameters that improved on those of Ptolemy.

Islam had no counterpart to the emerging universities of the Christian West, and we look in vain for an Islamic thinker of sufficient originality to challenge the foundations of Aristotelian/Ptolemaic cosmology. Nevertheless, discussions of *shukuk*, or doubts, concerning Ptolemy were appearing regularly by the 10th century. The most obvious target was the Ptolemaic equant, which violated the basic principle of uniform circular motion, but the epicycle and eccentre also came in for criticism because they involved motions that, while uniform, did not take place about the central Earth. A philosophical purist in this was the Andalusian Muhammad ibn Rushd (1126–98), known to the Latins as Averroes; in the West Aristotle was to become 'The Philosopher', and Averroes 'The Commentator'. Averroes recognized that Ptolemaic models 'saved the appearances' – reproduced the observed motions of the planets – but this did not make them true. His contemporary and fellow Andalusian Abu Ishaq al-Bitruji (Alpetragius) attempted to devise alternative models that met Aristotelian requirements, but of course with very unsatisfactory results.

In Cairo Ibn al-Haytham (Alhazen, 965–c.1040) tried to adapt the Ptolemaic models so that they could take on physical reality. In his *On the Configuration of the World* the heavens were formed of concentric spherical shells, within whose thicknesses smaller shells and spheres were located. His work was translated into Latin in the 13th century, and was to become one of the influences on Georg Peurbach in the 15th century.

The equant had long aroused misgivings among even the most practical-minded astronomers, and at Maragha in the 13th century al-Tusi succeeded in devising a geometrical substitute involving two small epicycles; Copernicus was at one stage in his career to adopt a similar device, and for the same reason, though historians have not yet identified an unambiguous link between them. An attempt to devise planetary models that were purged of all objectionable elements was made by Ibn al-Shatir, *muwaqqit* of the Umayyad mosque at Damascus, in the middle of the 14th century. His lunar model avoided the huge variations in the apparent size of the Moon implied by the lunar model of the *Almagest*, his solar model was based on new observations, and all his models were free not only of equants but also of eccentres. Epicycles, however, he found unavoidable, for reasons that we can well understand. By the time of al-Shatir, however, the Latin West had developed its own astronomical tradition, and was no longer reliant upon translation from the Arabic.

This independence had been slow in coming. In the Roman world, Greek had continued to be the language of scholars, and none of the major astronomical works of antiquity was written in Latin. With the collapse of the Roman Empire, knowledge of Greek almost completely disappeared in the West, so that the classics of ancient astronomy – even if available – could no longer be read. Ancius Manilius Severinus Boethius (c.480–524/5), a high official in the Roman Gothic kingdom, set himself to translate into Latin as many treatises of Plato and Aristotle as possible, but he had already left it too late. However, before his execution for defiance of his king over an injustice, Boethius did manage to translate a number of Greek works, several of them on logic, and these he set alongside logical writings by Roman authors such as Cicero. In this way he bequeathed to later centuries a corpus of texts in what became the one secular area of study where the medieval student might 'compare and contrast' and so come to his own conclusions. As a result, logical consistency was to become an obsession in the medieval university, where debates over the validity of epicycles, or whether certainty could in principle be attained in a planetary model, were meat and drink to the young students in Arts.

Just one (incomplete) work of Plato made its way into Latin during this period: his cosmological myth, *Timaeus*, two-thirds of which was translated by Calcidius (in the 4th or 5th century), who supplied a lengthy commentary. Astronomical works written in Latin in the early Middle Ages make sad reading, although the basic fact of the sphericity of the Earth was never lost to sight. Ambrosius Theodosius Macrobius, an African who lived in the early 5th century, wrote a commentary on Cicero's *Dream of Scipio*, and in this he expounded a cosmology in which a spherical Earth lay at the centre of the sphere of stars, which rotated daily from east to west. As it did so, it dragged the planetary spheres with it, though each of these also had its individual motion in the opposite direction. Macrobius is vague about the order of the planets because his sources differed. Martianus Capella of Carthage (*c.*365–440) wrote *The Nuptials of Philology and Mercury*, an allegory of a heavenly marriage at which each of the seven bridesmaids presented a compendium of one of the Liberal Arts. This account of astronomy is notable for the explanation of why Venus and Mercury are always seen near the Sun in the sky: they are circling the Sun, and so they accompany the Sun as it circles the Earth.

Christianity, like Islam, presented challenges to astronomers, chief among them the calculation of the date of Easter. In simple terms, Easter Day is the Sunday that follows the full moon that follows the spring equinox, and so its date in any given year depends on the cycles of both Sun and Moon. It might have been possible for the Christians of Alexandria, as inheritors of the accurate values for month and year handed down from Babylon, to calculate the appropriate date of Easter for some years ahead; but the Church authorities took the more practical course of trying to identify a period consisting of a number of years that almost equalled an integral number of months, and establishing the dates of Easter within the approaching years of this period. Once established, such a sequence could be repeated for future periods, indefinitely.

The cycle eventually adopted was one discovered by Babylonian astronomers in the 5th century BC but credited to the Greek Meton, whereby 235 lunar months equal 19 years (with an error of only a couple of hours). The definitive treatise, *On the Divisions of Time*, was written in 725 by the Venerable Bede (672/673–735) of Jarrow in England. In the calendar laid down by Julius Caesar, a leap year occurred every fourth year (without exception); every four years, therefore, the day of the week on which a given date occurred advanced by five, and so in $7 \times 4 = 28$ years it would return to its original day. Bede combined this with the 19-year Metonic cycle to produce an overall cycle of $19 \times 28 = 532$ years that catered for the luni-solar pattern of Easter together with the requirement that it occur on a Sunday.

The revival of astronomy – and astrology – among the Latins was stimulated around the end of the first millennium when the astrolabe entered the West from Islamic Spain. Astrology in those days had a rational basis rooted in the Aristotelian analogy between the microcosm – the individual living body – and the macrocosm, the cosmos as a whole. Medical students were taught how to track the planets, so that they would know when the time was favourable for treating the corresponding organs in their patients.

In 1085 the great Muslim centre of Toledo fell into Christian hands, and the intellectual riches of Islam and, more especially, Greece became accessible. Translators descended on Spain, the most notable being Gerard of Cremona (*c*.1114–87), whose innumerable translations included the *Almagest* and the Toledan Tables of al-Zarqali (d. 1100). These tables were then adapted for other places and proved immensely successful, although the underlying planetary models remained a mystery for the time being.

If the 12th century was the era of translations, the 13th was that of the assimilation of the works translated. In the emerging universities, Latin was the *lingua franca*, and so there was no language barrier to prevent students and teachers going where they wished. Prospective lawyers might go to Bologna and medical students to Padua, but in most disciplines Paris was pre-eminent.

There, as elsewhere, the Faculty of Arts provided the basic education in literacy and numeracy, through the medium of the seven Liberal Arts, which included astronomy. The Arts students were mostly boys in their teens, and the invention of printing lay in the future, so the level of instruction was inevitably elementary. A minority of students would eventually stay on for theological, medical, or legal studies in one of the higher faculties; medicine and law enjoyed their traditional prestige, while the writings of Augustine and the other Fathers of the Church ensured that theology was a challenging intellectual discipline. There was therefore tension between the teachers in these higher faculties and those trapped in the humdrum routine of Arts.

The bulk of the new translations, however, belonged to Arts, and provided the Parisian Masters of Arts with a lever to use in their struggle for improved status. At the same time, the arrival of the Aristotelian corpus, which owed nothing to Christian Revelation and which seemed to challenge certain basic Christian doctrines, aroused misgivings among the theologians. There followed decades of turmoil at Paris, until a synthesis was achieved by the Dominican friar Thomas Aquinas (1225–74), who assimilated Aristotle into Christian teaching so successfully that the 17th century would find it hard to make a separation of the two.

Research was not then the function of a university, and in astronomy the immediate teaching need had been for an elementary textbook that the young students might use. An attempt at this was made in the mid-13th century by John of Holywood (Johannis de Sacrobosco), but his *Sphere* was hopelessly inadequate when faced with the challenge of explaining the motions of the Sun, Moon, and lesser planets. Nevertheless, after the invention of printing the work offered more competent astronomers an excuse to write elaborate commentaries, and in this form it would become one of the best-sellers of all time.

Later in the 13th century an anonymous author made good some of the defects of the *Sphere* with his *Theory of the Planets*. This gave a simple (if only partly satisfactory) account of the Ptolemaic models of the various planets, with clear definitions. Meanwhile, at the court of King Alfonso X of Castile, the old Toledan Tables were replaced by the Alfonsine Tables; modern computer analysis has shown that these tables, which would be standard for the next 300 years, were calculated on Ptolemaic models with only the occasional updating of parameters.

It was not until the 14th century that the Latin West had sufficiently mastered its heritage from the past to be able to break new ground. One development of significance for astronomy came in terrestrial physics, for it was by arguing from the motion of projectiles that Aristotle had most convincingly demonstrated the Earth to be at rest: an arrow fired vertically fell to the ground at the very place from which it had been fired, and this proved that the Earth had not moved while the arrow was in flight.

However, Aristotle was at his least convincing when discussing the physics of projectile motion. An earthly body such as an arrow, he argued, would naturally move downwards towards the centre of the Earth, and its upward (and therefore unnatural) motion must be imposed upon it by an outside force – and not only imposed, but maintained for as long as the arrow was climbing. Aristotle thought the air itself was the only agent available to maintain the upward motion of the arrow; but sceptics had pointed out that this was implausible, since it was possible to fire arrows upwards in the teeth of a gale.

The Parisian masters Jean Buridan (*c.*1295 – *c.*1358) and Nicole Oresme (*c.*1320–82) agreed with Aristotle that a force must be at work, but they rejected any role for the air in this. They argued that an 'incorporeal motive force' must be imposed by the archer on the arrow, a force they termed 'impetus'. Buridan suggested that the heavenly spheres – which though frictionless needed a permanent motive force (angelic intelligences?) if they were to rotate for ever – would spin eternally if endowed with the motive force of impetus at the Creation.

Oresme saw a significant implication of the concept of impetus. *If* the Earth were indeed rotating, the archer as he stood on its surface would be moving with it. As a result, as he prepared to fire the arrow, he would unknowingly confer on the arrow a sideways impetus. Endowed with this impetus, the arrow in flight would travel horizontally as well as vertically, keeping pace with the Earth, and so would fall to ground at the very place from which it had been fired. The flight of arrows, he said, therefore contributed nothing to disputes as to whether the Earth was or was not at rest. Nor, for that matter, did the other arguments traditionally invoked, including those from Scripture. Oresme was of the opinion that the Earth was indeed at rest; but it was no more than an opinion.

The invention of printing in the 15th century had many consequences, none more significant than the stimulus it gave to the mathematical sciences. All scribes, being human, made occasional errors in preparing a copy of a manuscript. These errors would often be transmitted to copies of the copy. But if the works were literary and the later copyists attended to the meaning of the text, they might recognize and correct many of the errors introduced by their predecessors. Such control could rarely be exercised by copyists required to reproduce texts with significant numbers of mathematical symbols. As a result, a formidable challenge faced the medieval student of a mathematical or astronomical treatise, for it was available to him only in a manuscript copy that had inevitably become corrupt in transmission.

After the introduction of printing, all this changed. The author or translator could now check proofs and ensure that the version set in type was faithful to his intentions; and the printer could then multiply perfect copies, to be distributed throughout Europe and available for purchase at prices that were modest compared with the cost of a handwritten manuscript.

Within a few decades the achievements of the Greek astronomers had been mastered and indeed surpassed. The *New Theories of the Planets* of Georg Peurbach (1423–61), the Austrian court astrologer, which appeared in print in 1474, described the Ptolemaic models that underlay the Alfonsine Tables. It also described physically-real representations of these same models, and it may have been the shortcomings of these that led Copernicus to take up astronomy.

In 1460 Peurbach and his young collaborator, Johannes Müller (1436–76) of Königsberg (in Latin, Regiomontanus), met the distinguished Constantinople-born Cardinal Johannes Bessarion (*c.*1395–1472). Bessarion was anxious to see the contents of *Almagest* made more accessible, and he persuaded the two astronomers to undertake the task. Peurbach died the following year, but Regiomontanus completed the assignment. Their *Epitome of the Almagest*, half the length of the original, appeared in print in 1496. It remains one of the best introductions to Ptolemy's masterpiece. The *Almagest* itself was published in an obsolete Latin translation in 1515, in a new translation in 1528, and in the Greek original in 1538. In 1543 a book would appear that surpassed it.

Nicolaus Copernicus (1473–1543) was born in Torún, Poland, and studied at the University of Cracow, where the professors of astronomy had made no secret of their dissatisfaction with the concept of the equant. He then went to Italy, where he studied canon law and medicine, learned some Greek, and developed his interest in astronomy. He is said to have lectured on the subject in Rome around 1500, to a large audience. In 1503 he returned to Poland to take up an administrative canonry of Frombork where his uncle was bishop, and he remained in the diocese for the rest of his life.

Whereas voluminous works by Aristotle were available in Latin in the later Middle Ages, Plato had fared less well: only two minor dialogues had been added to the *Timaeus* that Calcidius had translated in part long ago. All this changed in the Renaissance, as close contacts were resumed with the Greek world, resulting in an influx of Greek scholars into the West prior to the sack of Constantinople in 1453. Plato's dialogues were recovered and admired for their literary qualities, and his mathematical outlook on the cosmos began to supplant that of Aristotle the naturalist. Astronomers began to look for harmony and commensurability in planetary theory, and they failed to find it in the models of Ptolemy, even though the Alfonsine Tables continued to meet the need for tables of reasonable accuracy. In particular, the equant was seen as 'a relation that nature abhors', in the words of Copernicus's disciple Georg Joachim Rheticus (1514–74).

It happened that by now Ptolemy's *Planetary Hypotheses* had been lost to sight, and with it his overall cosmology. The *Almagest* offered models for the individual planets, but Ptolemy's apparent failure to present an integrated picture of the cosmos meant, as Copernicus would put it, that past astronomers

> have not been able to discover or to infer the chief point of all – the structure of the universe and the true symmetry of its parts. But they are just like someone taking from different places hands, feet, head, and the other limbs, no doubt depicted very well but not modelled

from the same body and not matching one another – so that such parts would produce a monster rather than a man.

It was from aesthetic considerations such as these, as much as from specific problems like the absurd variations in the apparent size of the Moon on the Ptolemaic model, that pressure for reform developed – even though in themselves the Ptolemaic models (with improved parameters) did all that could reasonably be asked of them.

There were clues as to the direction a reform might take. Aristotle's custom of citing those whom he intended to refute meant that every student knew of ancient authors who had argued that the Earth was in motion – and Aristotle's rebuttal was no longer so convincing. Peurbach had remarked on how for some unknown reason an annual period cropped up in the model of every single planet. Whatever it was that set Copernicus thinking, not many years after his return from Italy a manuscript by him entitled *Commentariolus*, or *Little Commentary*, began to circulate. In it he outlined his dissatisfaction with existing planetary models, their equants coming in for special criticism. He proposed a Sun-centred alternative, in which the Earth became a planet orbiting the Sun with an annual period, while the Moon lost its planetary status and became a satellite of Earth.

He showed how this led at last to an unambiguous order for the planets (now six in number), in both period and distance. We saw how Ptolemy plausibly assumed that the slower-moving planets were the highest in the sky; but this did not settle the order of Sun, Mercury, and Venus, which accompany each other as they move against the background stars and therefore appear to have the same period of one year. Once this period was accepted as being in fact that of the Earth-based observer, the true periods of Mercury and Venus could be identified, quite different from each other and from that of the Earth; and so an unambiguous sequence of periods could be established.

Copernicus was also able to measure the radii of the planetary orbits as multiples of the Earth–Sun distance; for example, when Venus appears furthest from the Sun (at 'maximum elongation'), the angle Earth–Venus–Sun is a right angle, and by measuring the angle Venus–Earth–Sun the observer can establish the shape of the triangle and therefore the ratios of its sides. The sequence of periods and the sequence of distances proved to be identical. He was later to say of this:

> Therefore in this arrangement we find that the world has a wonderful commensurability, and that there is a sure linking together in harmony of the movement and magnitude of the orbital circles, such as cannot be found in any other way.

This was a powerful consideration in an age dominated by the Platonic search for harmony in the cosmos. Meanwhile, at the more detailed level, Copernicus made a start in *Commentariolus* on developing equant-free models for the planets and Moon.

Years passed, during which Copernicus developed his mathematical astronomy, remote from the intellectual centres of Europe. In 1539, Rheticus, then a teacher of mathematics in the University of Wittenberg, paid him a visit. He found himself enthralled with Copernicus's achievement in developing geometrical models of planetary motions that rivalled those of the *Almagest*, but which were incorporated into a coherent, Sun-centred cosmovision. He secured Copernicus's permission to publish a *First Report* on this work, which he did the following year. He also persuaded Copernicus to allow him to take the completed manuscript (known by its abbreviated Latin title of *De revolutionibus*) to Nuremberg for printing. The task of seeing the work through the press he delegated to a Lutheran clergyman, Andreas Osiander (1498–1552), who with the best of intentions inserted an unauthorized and unsigned preface to the effect that the motion of the Sun was proposed merely as a convenient calculating device; the result was that readers who got no further than this preface had no inkling of the author's true purpose.

The overwhelming bulk of Copernicus's book was concerned with (equant-free) geometrical models of the planetary orbits, and in their daunting complexity these matched the models of the *Almagest*. Proof that they could become the basis of accurate planetary tables – demonstration that the heliocentric approach could pass the practical test – was to come later, with the publication in 1551 of the Prutenic Tables of Erasmus Reinhold (1511–53), which were computed using Copernicus's models. It was in the brief Book I of *De revolutionibus* that Copernicus outlined the striking consequences that follow from the basic assumption that the Earth is an ordinary planet orbiting the Sun.

As we have seen, the list of planets ordered by period was identical with the list of planets ordered by distance. Equally striking, the mysterious 'wanderings' that had given the planets their name now became the obvious and expected consequence of observing one planet from another: Mars seems to go backwards when opposite to the Sun in the sky simply because it is then that the Earth overtakes it 'on the inside'. There is no longer any mystery as to why two planets, Mercury and Venus, are always seen near the Sun, whereas the other three may be observed at midnight; the orbits of Mercury and Venus are inside that of the Earth, while the others are outside.

True, the Earth was anomalous in being the only planet to have a satellite. It was true, too, that the 'fixed' stars did not show the apparent annual motion that one would expect if they were being observed from an Earth in annual orbit (Copernicans retorted that the stars were far away, and their 'annual parallax' was therefore too small for observers to detect). But these were details. The heliocentric universe was a true cosmos:

> In the centre of all resides the Sun. For in this most beautiful temple, who would place this lamp in another or better place than that from

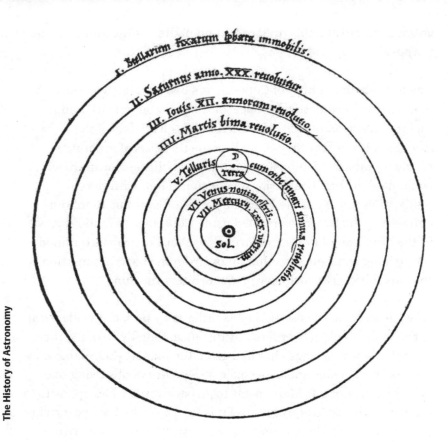

**11.  The outline diagram of the solar system, from Book I of Copernicus's *De revolutionibus*, showing the planets with their approximate periods. Note that no. V, the Earth, is unique in having a satellite, the Moon. Galileo's later telescopic discovery that Jupiter also has satellites helped ease the embarrassment felt by Copernicans because of this anomaly.**

which it can illuminate the whole at one and the same time? As a matter of fact, not inappropriately do some call it the lantern of the universe; others, its mind; and others still, its ruler. The Thrice-Great Hermes calls it a 'visible god'; Sophocles's Electra, 'that which gazes upon all things'. And thus the Sun, as if seated on a kingly throne, governs the family of planets that wheel around it.

*De revolutionibus* is the culmination of the Greek programme to 'save the appearances' of the mysterious planets by geometrical models using combinations of circles rotating with uniform motion. It is an *Almagest* purged of equants, though every bit as complex. It would be a generation before its revolutionary implications sank in.

# Chapter 4
## Astronomy transformed

Copernicus may have been traditional in his aims and the way he set about achieving them, but his claim that the Earth is in motion had posed a whole range of problems. What moves the Earth? How is it that we have no sensation of movement, and that arrows fired vertically upwards fall to the ground at the place from which they were fired? If we observe the stars as we orbit every six months from one side of the Sun to the other, why can we not detect in them an apparent movement of 'annual parallax'? And how can we explain those passages of Scripture that appear to imply that the Sun is in motion?

Some were misled by the unsigned preface to *De revolutionibus* into believing that Copernicus himself made no claim that the Earth was truly in orbit; merely, that he could 'save the appearances' more successfully by using geometrical models in which the Earth was imagined as moving. Others – including nearly all the competent mathematical astronomers of the next generation – were preoccupied with exploiting these models precisely in order to save the appearances, and neglected the cosmological Book I in which Copernicus's beliefs were clearly stated. Others again looked for some sort of compromise, and among them was a man who was conservative where Copernicus was innovative, and innovative where Copernicus was conservative: Tycho Brahe (1546–1601).

Tycho was born into a noble Danish family, but instead of adopting the pattern of life expected of members of his class in a feudal society, he followed his academic inclinations, among which astronomy figured prominently. In 1563 there occurred a conjunction of Jupiter and Saturn. Because these are the slowest moving of the five planets, their conjunctions, when Jupiter overtakes Saturn, are rare, occurring only once every 20 years, and so are astrologically the most ominous. The teenage Tycho made observations around the time of the conjunction of 1563 and concluded that the date predicted for it in the 13th-century Alfonsine Tables was a month out, while even the modern Prutenic Tables based on Copernicus's models were a couple of days wrong. This, he decided, was unacceptable, and before long he committed himself to the reform of observational astronomy.

Copernicus, like his predecessors, had been content to work with observations handed down from the past, making new ones only when unavoidable and using instruments that left much to be desired. Tycho, whose work marks the watershed between observational astronomy ancient and modern, saw accuracy of observation as the foundation of all good theorizing. He dreamed of having an observatory where he could pursue the research and development of precision instrumentation, and where a skilled team of assistants would test the instruments even as they were compiling a treasury of observations. Exploiting his contacts at the highest level, Tycho persuaded King Frederick II of Denmark to grant him the fiefdom of the island of Hven, and there, between 1576 and 1580, he constructed Uraniborg ('Heavenly Castle'), the first scientific research institution of the modern era.

It had everything: four observing rooms, numerous bedrooms, dining rooms, library, alchemical laboratory, printing press. Elsewhere on the island, there was even a paper mill, so that Tycho was completely self-sufficient in the publication of his work. Four years later, Tycho enlarged his facilities with a satellite observatory,

12. The great mural quadrant of Tycho's observatory at Uraniborg. The wall was aligned north–south, and the observer (barely glimpsed at the extreme right) measured the altitude of a heavenly body as it transited the meridian. One assistant is shown calling out the time of the transit, while another is recording the observation. Tycho, and various features of the observatory, are portrayed in the painting on the wall.

Stjerneborg ('Castle of the Stars'), built partly underground so that its instruments – unlike those in Uraniborg – would be on stable mounts and sheltered from strong winds. But he was loath to admit any limitation in Uraniborg: a second observatory, he said, would prevent collusion between two teams of observers making parallel observations.

Tycho was to remain on Hven until 1597. By then Frederick had been succeeded by Christian IV, who had come to resent Tycho and his arrogant demeanour and who was making life increasingly difficult for him. Tycho therefore quit, and two years later became mathematician to the Emperor Rudolf II in Prague. Tycho had lost his enthusiasm for observations; some instruments remained on Hven and others were kept in store. But he had with him the vast collection of accurate observations of the Sun, Moon, and planets that his teams had made on Hven, and these were to prove decisive for the work of Kepler, who joined him in Prague as assistant and who succeeded him when he died in 1601.

Tycho was the first of the modern observers, and in his catalogue of 777 stars the positions of the brightest are accurate to a minute or so of arc; but he himself was probably most proud of his cosmology, which Galileo was not alone in seeing as a retrograde compromise. Tycho appreciated the advantages of heliocentic planetary models, but he was also conscious of the objections – dynamical, scriptural, astronomical – to the motion of the Earth. In particular, his inability to detect annual parallax even with his superb instrumentation implied that the Copernican excuse, that the stars were too far away for annual parallax to be detected, was now implausible in the extreme. The stars, he calculated, would have to be at least 700 times further away than Saturn for him to have failed for this reason, and such a vast, purposeless empty space between the planets and the stars made no sense.

He therefore looked for a cosmology that would have the geometrical advantages of the heliocentric models but would retain the Earth as the body physically at rest at the centre of the cosmos. The solution seems obvious in hindsight: make the Sun (and Moon) orbit the central Earth, and make the five planets into satellites of the Sun. But the path of discovery was, as so often, tortuous. By 1578 Tycho was thinking of making Venus and Mercury into satellites of the Sun, as had Martianus Capella a millennium before. By 1584 he would have turned all five planets into satellites, except that this would have implied that the sphere carrying Mars intersected with the spheres carrying the Sun.

It was then that he saw the implications of observations he had made back in the 1570s. In November 1572 a star-like object bright enough to be visible in the daytime had appeared in the constellation of Cassiopeia. The heavens (it was thought) had been changeless since the dawn of history, yet the object resembled a bright star. Although he was only 26 and Hven was still in the future, Tycho had already progressed as an observer to the point where he could be certain that the object was celestial rather than atmospheric. Others disputed this, but Tycho was to subject their observations to a critical analysis that eventually settled the issue: the heavens could indeed change.

It might be thought that comets were already ample proof of this; but for Aristotelians, the comets' 'coming to be and passing away' amply demonstrated their terrestrial – or, more exactly, atmospheric – nature. As Aristotle himself had explained, comets resulted from the effect of the rotating heavens on the air and fire that surround the Earth, 'so whenever the circular motion stirs this stuff up in any way, it bursts into flame at the point where it is most inflammable'.

As long as the heavens had been changeless, there had been little reason to dissent from Aristotle's claim that comets were atmospheric. But, in the aftermath of the new star, Tycho harboured doubts. If only Nature would provide him with a bright comet, he would measure its height and establish whether it was atmospheric or celestial. In 1577, when Uraniborg was under construction, Nature obliged; and Tycho established that the comet was moving freely among the planets. As he later realized, this showed that the Earth-centred spheres thought to carry the planets did not exist.

With this, the physical objection to his cosmology disappeared, and in his 1588 book on the comet he presented his system in outline, along with detailed geometrical models for the motions of the Sun and Moon.

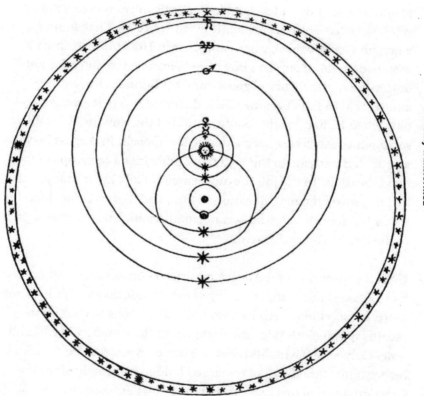

13. **The Tychonic system in outline. The Earth is at the centre, and around it orbit the Moon and the Sun. The Sun is itself orbited by five planets, which it carries around the Earth. The stars are immediately beyond the outermost planet, Saturn. The relative motions are the same as in the Copernican system, and this created great difficulties for Galileo in his campaign in favour of the Copernican theory.**

The stars lay immediately beyond the realm of Saturn, at a distance of some 14,000 Earth radii, so that the Tychonic universe was even more compact than the Ptolemaic.

A number of similar compromises were floated in the decades to come, and many found them attractive. They were to infuriate Galileo Galilei (1564–1642) in his campaign in support of Copernicus because they were so difficult to refute. As professor of mathematics at Padua in the 1590s, Galileo had used the daily and annual motion of the Earth in an attempt to explain the puzzling phenomenon of the tides, but his Copernicanism was less than whole-hearted until the dramatic events of 1609. That summer, when he was in Venice, word came that in Holland instruments with two pieces of curved glass were being used to make distant objects seem near. Curved glass was a traditional source of amusement in fairs because of the distorted image it created, and only when he had reliable confirmation of the rumour did Galileo set about constructing such a device for himself. In August he was able to demonstrate to the Venetian authorities a telescope of 8× magnification, 'to the infinite amazement of all'. Later that year he had improved the magnification to 20×, and it is no coincidence that a few months later he became mathematician and philosopher to the Grand Duke of Tuscany.

Until the invention of the telescope, each generation of astronomers had looked at much the same sky as their predecessors. If they knew more, it was chiefly because they had more books to read, more records to mine. All this now changed. In the coming months and years Galileo saw with his telescope wonders vouchsafed to no one before him: stars that had remained hidden from sight since the Creation, four moons that orbited the planet Jupiter, strange appendages to Saturn that would be recognized as rings only half a century later, moon-like phases of Venus, mountains on the Moon not very different from those on Earth, even spots on the supposedly perfect Sun. He was able to confirm the suggestion of Aristotle that the Milky Way is composed of myriads of tiny stars.

He found that the disc-like shape that stars seem to have when viewed with the naked eye was an optical illusion, so that if Copernicans found themselves forced to banish them to remote regions to avoid the difficulty over annual parallax, they did not thereby have to make them physically huge. Galileo could say of his predecessors, 'If they had seen what we see, they would have judged as we judge'; and ever since his time, the astronomers of each generation have had an automatic advantage over their predecessors, because they possess apparatus that allows them access to objects unseen, unknown, and therefore unstudied in the past.

Galileo lived in an age before scientific journals provided a forum for rapid dissemination of new work, and when leisurely publication in book form was the norm. But his telescopic discoveries could not wait, and he was able to announce the earliest of them within months, in his brief *Starry Messenger* (1610). This was followed in 1613 by *Letters on Sunspots*. Several of his revelations supported Copernicus, none more so than the moon-like sequence of the phases of Venus, which he announced in the *Letters*.

In the Ptolemaic system, Venus was below the Sun in the sequence of planets; furthermore, to 'save' the fact that Venus never appears far from the Sun in the sky, Ptolemy's model required the centre of Venus's epicycle to be on the straight line from the Earth to the Sun. As a result, in the model, Venus always lay somewhere between the Earth and the Sun.

What then if Venus proved to be a dark body illuminated by the Sun? If the Ptolemaic model was correct, the illuminated half would always face partly away from Earth, and so the planet would never appear to us as a circle of light, like the Moon when full. For Copernicus, by contrast, Venus orbited the Sun on a path inside that of the Earth. When near to Earth it would appear to have a crescent shape because its illuminated half would be facing away from Earth, as in the Ptolemaic model; but when away on the far side of the Sun it would appear full.

And this was just what Galileo had witnessed. Ptolemy was wrong, disproved by a decisive observational test. Copernicus was therefore right – or so Galileo would have us believe. But the phases of Venus tell us only about the *relative* positions of Sun, planet, and Earth, and nothing about which of the three is physically at rest; and the relative motions are broadly the same in both the Copernican and Tychonic systems. The Tychonic system was therefore unscathed by Galileo's discovery.

To Galileo this was most unwelcome, because Ptolemy had long since been discarded by those who believed the Earth to be at rest in favour of the Tychonic, or one of the 'semi-Tychonic', systems. His telescopic discoveries had made him an ardent propagandist for Copernicus; but proving Ptolemy wrong was easier than proving Copernicus right. And so he continued to act as though the choice was still between Ptolemy and Copernicus: as late as 1632 he gave his Copernican manifesto the title of *Dialogue on the Two Great World Systems, the Ptolemaic and Copernican.*

To resolve the physical objections – how is it that we Earthlings are hurtling through space, deluded all the while into believing we are in fact on *terra firma*? – Galileo created a new conception of motion. Motion – change of all kinds, of which change of place is only one – had been fundamental to Aristotelian philosophy, for a natural body expressed its nature by how it behaved, how it moved. In Aristotle's view, motion demanded explanation, rest did not.

Galileo, to the contrary, set out a new way of looking at the world, in which it was change of motion – acceleration – that called for explanation, while steady motion (of which rest was now merely a special case) was a state that needed no explanation. He imagined a ball rolling on the surface of a perfectly smooth, spherical Earth, and saw no reason why the ball should ever come to rest: it would remain indefinitely in a state of uniform motion, rolling around the centre of the Earth. Similarly, the Earth itself orbited the centre of the solar system, in a state of uniform motion, which was why Earthlings were unaware of their movement.

Galileo had a gift for friendship but also a gift for enmity, and the view that the Earth moves had long been seen as in apparent contradiction to certain Scriptural phrases. In 1613 Galileo wrote a semi-public letter to a friend that was to become the basis of his *Letter to the Grand Duchess Christina* (written in 1615 but not published until 1636). This is now recognized as a classic statement of the traditional Catholic position – that the Bible teaches us how to go to heaven, not how the heavens go – but the period of the Counter-Reformation was no time for a layman to pronounce on the interpretation of Scripture. In 1614 a preacher mounted an attack on him by turning a text from the Acts of the Apostles into perhaps the best pun in ecclesiastical history: 'Ye men of Galilee, why stand you gazing up to heaven?' Galileo, ignoring warnings from friends, stood his ground; the dispute escalated, and the Vatican became involved. Eventually, in February 1616, Galileo was interviewed by the saintly Cardinal Robert Bellarmine. Bellarmine, who was prepared to revise his traditional position on the stability of the Earth but only when compelling proof was forthcoming, notified Galileo privately that he was no longer permitted to believe that the Copernican system was true, or to defend this belief.

Time passed, and in 1623 the election of a new Pope who had been Galileo's friend and supporter encouraged him to resume his campaign in support of Copernicanism, and this led eventually to the publication of his *Dialogue* in 1632. Exactly what it was that so upset the Roman authorities is still debated, but the outcome was that Galileo was condemned to house-arrest. Comfortable though this in fact proved to be, his condemnation was a setback to astronomy in Catholic countries, and the many Jesuit astronomers found themselves expected to support the Tychonic system or a similar compromise.

There was in Galileo's makeup a certain laziness, in particular a reluctance to engage in hard mathematics, and this cost him dear in his campaign as Copernican propagandist; for he remained oblivious to the service rendered to the Copernican cause by a contemporary who saw the planets as bodies driven round by forces emanating from the massive, central Sun – and who was thereby transforming astronomy from applied geometry to a branch of physics, from kinematics to dynamics. Johannes Kepler (1571–1630) was born at Weil der Stadt near Stuttgart and studied at the University of Tübingen, where his fair-minded teacher in astronomy set out the pros and cons of the various cosmologies on offer, including that of Copernicus. Kepler then commenced studies in theology, but in 1593 the authorities nominated him to teach mathematics in Graz, and with reluctance he complied.

Settled in Graz, he began to puzzle over the structure of the universe, which he saw as created by God, the great geometer. Copernicus, he believed, had discovered the basic layout of the universe, but not the reasons that had motivated God in His selection of this particular universe from among the possible options. In particular, why had God created just six planets, and why had He given the five spaces between them the sizes that they have? Eventually it occurred to Kepler that the number of spaces equalled the number of the regular solids (pyramid, cube, and so on), shapes that must have a profound appeal for any geometer, human or divine. And so Kepler investigated the geometry of nests of six concentric spheres, each pair separated by one of the five regular solids in such a way that the inner sphere of the pair touched the planes of the solid while the outer sphere passed through its vertices; and eventually he identified a particular nest that reproduced, to a reasonable approximation, the radii as calculated by Copernicus.

This did not address the question of the speeds of the planets, and Kepler took the epoch-making step of approaching this problem in

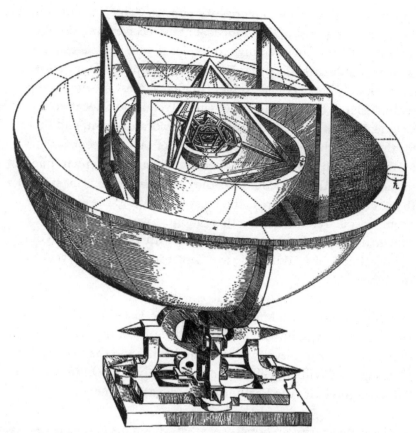

**14.** **The geometrical relationships embodied by God in His universe, according to Kepler's *Cosmographic Mystery* (1596)**

terms of the physical influence of the central – and massive – Sun. After all, Copernicus had shown that the further a planet was from the Sun, the slower it moved in its orbit. Perhaps this was because the Sun was the cause of the planets' movement, and the effectiveness of this cause diminished with increasing distance.

It would be more than two decades before Kepler would establish the actual pattern of the planetary speeds, but his early speculations, published in *Cosmographic Mystery* (1596), served notice on the astronomical community that a new talent had appeared on the scene. Kepler sent a copy to Galileo, urging him to come out in support of Copernicus, but he received only a polite response. Tycho, however, invited Kepler to join him on Hven, notwithstanding the fact that Kepler's book was the first that was irredeemably heliocentric. Kepler decided against moving to this remote island; but early in 1600, when Tycho had relocated to Prague, Kepler decided to pay him an exploratory visit, and there for three months he worked on the orbit of Mars. Except for Mercury, which is hard to track because it spends so much time lost in the glare of the Sun, Mars has the orbit that departs most from the circular and therefore is the most difficult to 'save' with a geometry of circles. After the three months were up, Kepler returned to Graz, but he was soon back with Tycho, who had only a short time to live. Within a year Kepler found himself Tycho's successor.

Kepler's 'warfare' with Mars, the god of war, lasted for several years. His campaign was, he said, based on the Sun-centred vision of Copernicus, the observations of Tycho Brahe, and the magnetical philosophy of William Gilbert (1544–1603), whose *On the Magnet* (1600) argued that the Earth itself was a vast magnet.

Kepler is that rarity, an honest scientist who in his publications does not launder his account of his research to make the path to his conclusions seem direct and untroubled. Kepler requires his reader to follow him in the maze of calculations, and he is no fit subject for any sort of Introduction, let alone a Very Short one. But the essential point is clear: Kepler abandoned the geometrical models of traditional astronomy, the study of *how* planets move, in favour of physics, the study of what forces cause them to move as they do.

This shift from kinematics to dynamics had been almost inevitable, once Tycho had demonstrated the non-existence of the heavenly spheres commonly thought to carry the planets round. Why these spheres continued to spin had been of minor interest – most held that angelic intelligences drove them on, Buridan had postulated an impetus given to each sphere at the Creation, while Copernicus thought that all natural spheres naturally spun. But take away the planetary spheres, and you are left with planets that are isolated bodies in orbit. What could it be that drives them round, almost in the manner of projectiles? Once this question was given centre-stage, the success of the heliocentric hypothesis was assured: it made dynamic sense for the relatively small Earth to orbit the massive Sun, but not vice versa.

In 1609 Kepler published his solution to the problem of Mars, in a work whose challenging title proclaimed the reorientation of astronomy: *New Astronomy Based upon Causes, or Celestial Physics, Treated by Means of Commentaries on the Motions of Star Mars*. Early on, Kepler had come to understand that he must take the real, physical Sun as the hub of his solar system, and not some point that was geometrically convenient. Equally, he must give a combined account of the motions of the planet in longitude and latitude; it was no longer acceptable to devise two geometrical models, each doing the job for one or other coordinate, but incompatible with each other.

Tycho's data were numerous enough to provide generous amounts of information about Sun, Earth, and Mars when they were in special configurations. Thus when Mars was exactly opposite to the Sun in the sky, an Earth-based observation could do duty for an observation from the Sun at the centre of the system; and Tycho had records of a number of these. They were accurate: when Kepler arrived at a model with a circular orbit that 'saved the appearances' to as little as 8 minutes of arc (good enough to match the observations of any earlier observer), Kepler knew that because Tycho's accuracy was much better than 8 minutes, the model was not satisfactory and he must reject it. Yet Tycho's observations were not (so to say) *too* accurate: the orbit of the real Mars is disturbed by the pulls of other planets, and so hypothetical observations of perfect accuracy would have prevented Kepler from reaching his eponymous laws.

Gilbert had argued that the Earth is a large magnet; perhaps the Sun was an even greater one. As the planets all orbit the Sun in the same direction, and move less urgently the further they are from the Sun, Kepler was led to think of the Sun as a rotating body that sends out into space a magnetic influence that pushes the planets round, the influence naturally being most effective on the nearest planets. Without this continual influence, Kepler believed, each planet would stop dead in its tracks – inertia in our everyday sense of the word.

But there was more to it than this, for the planetary orbits are not simple circles. The planets vary their distances from the Sun, and to account for this Kepler introduced into the Sun a second force that attracted the planet for part of its orbit and repelled it over the rest.

These physical intuitions guided Kepler in his analysis of Tycho's observations. As it happened, Kepler discovered his 'second law', which tells us about the speed of the planet in its orbit, before the 'first law', which tells us what the orbit is. According to the first law, the planet moves in an ellipse with the Sun at one focus. Ellipses had been well understood by geometers since the 2nd century BC, and it was a masterstroke to define a planetary orbit, not with a Ptolemaic-type model bristling with epicycles, eccentres, and equants, but through a single, very familiar curve. But in arriving at the law Kepler's physical intuition had at one stage proved something of a barrier – for an ellipse has symmetry about the minor axis, whereas Kepler's first law tells us that the orbit is highly unsymmetric about this axis, the Sun being at one focus and the other being 'empty'.

We know the second law in the variant form that Newton later found to be a *consequence* of his law of gravitational attraction: a line from the Sun to the planet traces out equal areas in equal times. Mathematicians were barely able to deal with such a bizarre formulation, and they preferred other variants that were mathematically tractable, and could hardly be distinguished observationally from the area law: that the planet moves with a speed inversely proportional to the Sun–planet distance; or that the planet moves with apparently uniform speed when viewed from the empty focus. In our earlier discussion of the Ptolemaic equant in a circle, we saw why its equivalent in an ellipse – uniform speed viewed from the empty focus – gives such a good approximation to the area law.

*New Astronomy* set out how Mars – and by implication the other planets – moved, and what caused it to do so. But what of the overall pattern of the planetary system, which Kepler had first tackled in *Cosmographic Mystery*? This was one of many themes of *The Harmony of the World* (1619), along with such topics as the music generated by the planets in their orbits. Copernicus had rejoiced to find that the further a planet was from the Sun, the longer it took to complete a circuit; Kepler could now announce the formula that ensured this: the square of the period of a planet is in a fixed ratio to the cube of the radius of its orbit.

Kepler was by now engaged in making his work accessible to a wider readership, setting it out in a user-friendly question and answer form. The title of his *Epitome of Copernican Astronomy*, which appeared piecemeal between 1618 and 1621, paid tribute to his ultimate source of inspiration; but Copernicus would have been baffled to discover that his geometrical astronomy had now migrated to become a branch of physics.

Planetary theories had always been subjected to the ultimate practical test: could they be used to generate accurate tables? Tycho's own interest in astronomy had been kindled by shortcomings in the Copernican Prutenic Tables, and when Tycho had first presented the young Kepler to the Emperor Rudolf, the Emperor commissioned Kepler to work alongside Tycho on tables on which astronomers might at last rely. In 1627, when Rudolf and Tycho were long dead, Kepler published the Rudolphine Tables. Kepler too was dead when the French astronomer Pierre Gassendi became the first observer in history to see a transit of Mercury across the face of the Sun. The forecast in the Rudolphine Tables was 30 times more accurate than that in the Prutenic: Kepler's elliptical astronomy had passed the test. Yet his physical intuition, of planets that would instantly come to halt without the incessant urging of the Sun, was wholly implausible, and the formulation of the second law – the crucial one that defines the varying velocity of the planet in its orbit – was confused and confusing. Kepler had replaced numerous circles with single ellipses, and he had encouraged astronomers to see their discipline as 'celestial physics'. But the true dynamics of the planetary system was as mysterious as ever.

# Chapter 5

# Astronomy in the age of Newton

The outlook of the later Middle Ages had been dominated by Aristotle, and that of the Renaissance by Plato. But in the period following, the 'mechanical philosophy', alternatively known as the 'corpuscular philosophy', became increasingly attractive. It had originated with the Greek atomists, who explained the different qualities we perceive in the bodies around us as being the ways our senses interpret the movements of unchanging particles; and this appealed to an age that found a refreshing clarity about explanations that used concepts like speeds and shapes, concepts that were mathematical at least in principle. Machinery was becoming ever more ingenious – witness the great clock in the cathedral in Strasbourg. But in these machines, complex effects were being produced by simple (and intelligible) means: matter – cogs, weights, and so forth – in motion.

God was now the great clockmaker, His creation hugely complex in structure, but intelligible simply as matter in motion. Galileo was one of many attracted by this revival of ancient ideas, but it was his younger contemporary, René Descartes (1596–1650), who carried the mechanical philosophy to its extreme. At the school at La Flèche his Jesuit teachers introduced him to Galileo's telescopic discoveries within months of their announcement; more importantly, they instilled in him a profound admiration for the certainty to be found in geometrical theorems. For Descartes, there was a vast gulf between the certainly true and what was only very probably true; and he decided that to bridge the gulf one must imitate the reasoning of the geometers. Furthermore, it was the geometers who had always had the correct understanding of space: the infinite, homogenous, undifferentiated space of Euclid was not the idealized abstraction it had been taken to be, but the space of the real world.

As a philosopher, Descartes was ruthless. Unlike Galileo, who could never mention Aristotle without getting involved in a squabble, Descartes dismissed Aristotle with contempt and set about creating his own philosophy. Until now, all discussion had referred back to Aristotle, favourably or otherwise; soon all discussion would be referred back to Descartes.

In the Cartesian universe there were no longer any privileged places, such as the centre of the Earth or the centre of the solar system, that were different from ordinary places because it was around them that motions took place. Analysing the fundamental concept of matter *per se*, Descartes rejected properties like colour or taste that belong to some matter but not to all, and ended by concluding that matter and space were in many respects identical. This being so, space without matter – a vacuum – was an impossibility: the world is a plenum. Furthermore, since space was uniform, so was matter. This meant that the differences we perceive between this material object and that are due entirely to how the (uniform) matter is moving in the two spaces involved: motions are everything when it comes to understanding the universe.

Because God lives in the eternal present, He conserves this particular matter in the motion that it has at this instant of time, travelling in a specific direction with a given speed: the law of rectilinear inertia. And just as He conserves the total space of the universe, so He conserves the total amount of motion in the universe. This allows us to establish the laws that govern the transfer of motion from one object to another.

Because the universe is packed with matter, rectilinear inertia is a tendency rather than a reality. In practice, matter can move only if the matter ahead of it (and the matter behind it) also moves. As a result, matter normally moves in vortices, or whirlpools; these act rather like a centrifuge, with some of the matter forcing its way towards the exterior, and other matter as a result being pushed towards the centre. The latter we see as self-luminous, and we perceive the great assemblages of this luminous matter as the Sun and stars. The Sun is therefore nothing more than our nearest star, and similar stars are scattered everywhere throughout the infinite universe.

The Sun is at the centre of a great solar vortex that carries the planets round; these planets tend to fly off at a tangent, but they are constrained into closed orbits by the rest of the matter in the vortex. Among the planets is the Earth, itself the centre of a lesser vortex that carries the Moon. The Moon is therefore transported in both the solar and the terrestrial vortices, and because of this Newton was to find its motion hard to calculate. But although Descartes was a mathematician, and wrote in a letter that 'My physics is nothing but geometry', *The Principles of Philosophy* (1644), in which he set out his ideas, consists only of words and has no mathematical equations. Words are vague and malleable, and *The Principles* proved so adaptable that it could explain almost everything, while predicting almost nothing. The book could be understood by the innumerate; and the world-picture it expounded was to prove immensely attractive to devotees of the Parisian *salons*.

In Oxford and Cambridge in the later decades of the century, Aristotle still officially held sway, though within the individual colleges enterprising tutors were able to introduce their charges to the novelties of Cartesian philosophy. London, however, was home to the Royal Society, founded in 1660, and the Fellows were heirs to the 'magnetical philosophy' of William Gilbert, who had so greatly influenced Kepler's thinking. A leading figure in the Society was John Wilkins (1614–72), who in 1640 had published a second edition of his *Discovery of a World in the Moone*, in which he argued that Moon travel was theoretically possible because the magnetic influence of the Earth diminished with height: 'it is probable, that this magneticall vigor dos remit of its degrees proportionally to its distance from the earth, which is the cause of it.'

Early in the 1660s, Robert Hooke (1635–1703), Curator of Experiments to the Royal Society, even carried out tests to see whether the pull of the Earth was less at the top of a cathedral than it was at ground level. The results were of course inconclusive; but in the years that followed, Hooke steadily generalized his thinking about magnetism as an explanation for what we see about us in the solar system. By 1674 he was able to set out the stage he had reached in three remarkable 'Suppositions':

> First, That all Coelestial Bodies whatsoever, have an attraction or gravitating power towards their own Centers, whereby they attract not only their own parts, and keep them from flying from them, as we may observe the Earth to do, but that they do also attract all the other Coelestial Bodies that are within the sphere of their activity . . . .

Hooke believed that all the bodies of the solar system attracted other bodies – more exactly, those 'within the sphere of their activity' – by a force identical to the gravity that held the parts of the Earth together. About rectilinear inertia he was admirably clear:

> The second supposition is this, That all bodies whatsoever that are put into a direct and simple motion, will so continue to move forward in a streight line, till they are by some other effectual powers deflected and bent into a Motion, describing a Circle, Ellipsis, or some other more compounded Curve Line.

> The third supposition is, That these attractive powers are so much

the more powerful in operating, by how much the nearer the body wrought upon is to their own Centers.

But did they vary inversely with the distance itself ($f \propto 1/r$), or with the distance squared ($f \propto 1/r^2$), or what? Hooke could not say; he regarded the answer as relatively unimportant, merely one of the loose ends to be left to mathematicians.

The inverse-square law was an obvious candidate, because the brightness of a heavenly body falls off with the square of the distance. But there was a more cogent reason. Analysis of the dynamics of circular motion – of a stone whirled around in a sling, for example – combined with Kepler's third law of planetary motion suggested that if the planets were all to move around the Sun in strictly circular orbits and at steady speeds, the entire pattern of these orbits could be explained as resulting from solar attraction that diminished with the square of the distance. Yet the true orbits of the planets were elliptical; could it be shown that elliptical orbits likewise result from an inverse-square law of attraction?

By 1684 opinion in London had hardened: the inverse-square law was the answer, but no one could handle the mathematics needed to demonstrate this. Might Isaac Newton (1642–1727), the highly gifted but secretive mathematics professor in Cambridge, be able to do so? Edmond Halley (c.1656–1742) plucked up his courage and bearded Newton in his den. What, he asked, would be the shape of the orbit of a planet attracted to the Sun under an inverse-square law? Newton unhesitatingly gave the reply Halley was hoping for: an ellipse.

Newton had entered Cambridge in 1661, and by the middle of the decade he too had successfully analysed the dynamics of strictly circular motion. But he was facing serious problems in trying to understand the planetary orbits. In the Cartesian universe (which Newton then accepted as a true representation of reality), the Moon was carried in both the solar and the terrestrial vortices, and this made mathematical analysis difficult. As to the planets, there were several variants of Kepler's second law in circulation, observationally almost identical but conceptually worlds apart. Newton himself tried working with a number of equant versions of the law. Eventually, however, he came across a book that set out the law in the area formulation that we know today.

In 1679 Newton was still struggling to make sense of vortices, and still confused in his analysis of the dynamics of orbital motion, when he received a letter from Hooke. Hooke was now Secretary of the Royal Society, and eager to involve the Cambridge mathematician in its activities. He invited Newton to consider the consequences of 'compounding the celestiall motions of the planets of a direct motion by the tangent [inertial motion] and an attractive motion towards the centrall body'. Hooke saw orbital motion, not as the outcome of a struggle between centrifugal and centripetal forces, but as the effect of an attractive force on motion that would otherwise continue in a straight line.

Hooke also told Newton about his attempt (see Chapter 6) to prove the motion of the Earth by measurement of annual parallax. In his reply, Newton offered a new approach to the traditional 'proof' that the Earth must be at rest, because arrows fired vertically fell to the ground at the place from which they had been launched – or, equivalently, a stone dropped from a tower fell to the ground at the base of the tower. Newton pointed out that the top of the tower was further than the base from the centre of the Earth; and that since the Earth was in fact spinning, the stone when still at the top of the tower was travelling horizontally more rapidly than the ground at its base. Therefore, he argued, since the stone would retain its horizontal velocity while falling, it would in fact strike the ground ahead of the tower. And he went on to discuss how the stone would continue to move, in the imaginary situation in which it had the power to pass through the Earth unimpeded. By doing this Newton converted a problem of free fall into one of orbital motion.

Hooke had the satisfaction of pointing out a mistake in Newton's analysis, though an imagined passage through the Earth was too fanciful for his taste. However, he did reveal to Newton his commitment to an inverse-square law: 'my supposition is that the Attraction always is in a duplicate proportion to the Distance from the Center Reciprocall.'

Newton's reaction to any hint of criticism was to withdraw into his shell, at the same time secretly devoting his energies to establishing the truth of the matter. Although Hooke's proposed scenario was, he thought, divorced from the real (Cartesian) universe which was crammed with matter, Newton pursued the mathematical analysis – and made the extraordinary discovery that Hooke's planets would move in ellipses with the Sun at a focus, and that the line from the Sun to the planet would trace out equal areas in equal times, exactly as Kepler had declared the real planets to do. Could it be that Hooke's was the real world – a world largely empty, in which isolated bodies somehow influenced each other across the intervening space by attraction – and Descartes's plenum the imaginary?

We know little of how Newton's thinking developed between 1679 and the visit from Halley in 1684, except for a confused exchange of letters between Newton and the Astronomer Royal, John Flamsteed (1646–1719), as to whether a comet seen approaching the Sun in November 1680 was identical with the comet seen leaving the Sun the following month (it was); and if so, what had happened to it meanwhile and why. Newton at one point suggested that it might have been a single comet that 'fetched a compass round the Sun' – that it had gone around the back of the Sun – and he may already have been thinking that the comet's path was the result of the attraction of the Sun; but we cannot be sure. Certainly the visit from the suitably deferential and tactful Halley encouraged Newton to promise him written proof that elliptical orbits would result from an inverse-square force of attraction residing in the Sun. The drafts grew and grew, and eventually resulted in *The Mathematical*

*Principles of Natural Philosophy* (1687), better known in its abbreviated Latin title of the *Principia*. Contemporaries recognized in the title a challenge and rebuke to the wordy and fanciful *Principles of Philosophy* of Descartes.

The first draft ran to a mere nine pages. It analysed the orbit of a body moving with inertial motion in empty space, under the influence of a pull to a 'centre'. Such a body would obey Kepler's area law. If the pull was inverse-square, then the orbit would be a conic section – an ellipse, parabola, or hyperbola. If bodies moved in elliptical orbits with the pull to the focus, then the orbits would obey Kepler's third law; and vice versa. All three of Kepler's laws (the second in 'area' form), which had been derived by their author from observations, with the help of a highly dubious dynamics, were now shown to be consequences of rectilinear motion under an inverse-square force.

Newton had not yet arrived at the point of seeing attraction as a mutual force between celestial bodies large and small, as Hooke had done much earlier; and this is curious, since his draft stated that Kepler's third law applied to the Galilean moons of Jupiter, and to the five moons of Saturn (Titan had been discovered by Christiaan Huygens in 1655, and Gian Domenico Cassini had since discovered four more). These moons were therefore attracted by their parent planet, and one wonders why, if Saturn pulled Titan, it did not also pull the Sun. Perhaps the same thought occurred to Newton, for in his next draft attraction was universal.

In Newton's mind, the Cartesian universe, crammed full of matter, in which bodies constantly impacted on each other, had now given place to one that was almost empty, and in which bodies moved with rectilinear inertia modified by the attractions of all the other bodies – attractions that somehow reached across empty space. Newton was justifiably aghast at the complexity of the mathematical challenges that resulted, not least in the study of the ways in which the Moon behaves under the competing pulls of

Earth and Sun. For their part, Continental mathematicians would be shocked at Newton's retrograde step in invoking a mysterious 'attraction', for which he offered no mechanism and which sounded for all the world like the reintroduction of the dubious 'sympathies' and other 'occult qualities' that the mechanical philosophy had only recently eradicated.

After two millennia of observation and analysis, the how and the why of planetary orbits were at last understood, even if the nature of the inverse-square force remained mysterious. But what of comets? Newton was now certain that the two comets seen late in 1680 were one and the same, and that it had indeed 'fetched a compass' around the Sun. Comets, he concluded, were part of the same universal pattern, and in the *Principia* he showed that their orbits were conic sections (though not necessarily ellipses) and that they too obeyed Kepler's area law. This opened up the possibility that a comet moving in an elongated ellipse would regularly return to the solar system.

Hooke had long since suspected that the Earth's pull on a falling stone was the same as its pull on celestial bodies, and the same thought now occurred to Newton. But to compare the pull of the Earth on a stone with its pull on the Moon, he faced a mathematical challenge. He would have to combine the pulls on the stone of all the bodies that make up the Earth, pulls operating over distances that ranged from a few feet to thousands of miles. Leaving aside the question of how a force could operate with equal effectiveness through thin air and through miles of rock and earth – some of his followers would admit this could happen only by the direct fiat of the Creator – Newton proved the remarkable theorem that the combined pulls equalled the pull of the entire Earth imagined as concentrated at its centre.

He could now compare the total pull of the Earth on the stone (effectively over a distance of one Earth radius) with the pull by which the Earth drew the Moon into a closed orbit (at a distance of

60 Earth radii). He found the ratio was indeed about $60^2$:1. Earth and sky obeyed the inverse-square law of attraction.

As the drafts of *Principia* multiplied, so too did the number of phenomena that at last found their explanation. The tides resulted from the difference between the effects on the land and on the seas of the attraction of Sun and Moon. The spinning Earth bulged at the equator and was flattened at the poles, and so was not strictly spherical; as a result, the attraction of Sun and Moon caused the Earth's axis to wobble and so generated the precession of the equinoxes first noticed by Hipparchus. Several 'inequalities', or irregularities, in the motion of the Moon had been detected – one by Ptolemy and others by Tycho – and these too Newton was able to explain, qualitatively if not quantitatively.

Our satellite is easy to observe, and the pulls on it are simple to state, but highly complex to analyse mathematically. Newton set the agenda for 18th-century mathematicians of the highest talent: to show that the observed lunar motions can be fully explained by the inverse-square law. The historical investigation of these increasingly successful attempts is not for the faint-hearted; fortunately for us it belongs to the history of applied mathematics rather than to the history of astronomy.

Newton was able to use the observed motions of the moons of Earth, Jupiter, and Saturn to calculate the masses of the parent planets, and he found that Jupiter and Saturn were huge compared to Earth – and, in all probability, to Mercury, Venus, and Mars. It seemed, therefore, that the two massive planets had been located at the outer reaches of the solar system, where their powerful gravitational attraction would do least harm to the stability of the solar system. In time, however, even this providential arrangement would begin to experience perturbations, and the system would then 'want a reformation': Providence would step in to restore the original order, and thereby demonstrate God's continuing care for mankind.

Some Continentals, notably Gottfried Wilhelm Leibniz (1646–1716), who agreed with Newton that God was the great clockmaker and saw the universe as a masterpiece of machinery, were scandalized that Newton thought God so poor a workman as to need to rectify his blunders by working miracles in this way. But for Newton this was all part of God's plan from the beginning; He had entered into a servicing contract with the universe, to demonstrate His continuing care for His Creation.

Other Continentals found the concept of attraction retrograde: true, Newton had used this supposed force to explain many movements, but the force itself was so far inexplicable. Could it perhaps be explained on Cartesian principles? The antiquated geometrical formulation of the theorems of Newton's *Principia* had been enough to deter all but a few would-be readers. It was only when popularizations began to appear in the early decades of the 18th century, and more especially when Continental mathematicians successfully exploited the Newtonian programme and explained more and more aspects of the complex behaviour of the Moon, that the merits of attraction become indisputable. Any remaining doubts were dispelled in 1759 by the reappearance of a comet.

According to Cartesian physics a comet was a dead star whose own vortex had collapsed, and which then wandered from one vortex to another, although if it penetrated far enough into a vortex it might remain there as a planet. Newton, however, had claimed that comets obeyed Kepler's laws (in their generalized form), and that a comet whose orbit was an elongated ellipse would regularly reappear. Halley therefore searched the historical records, looking for three or more comets all with similar orbital characteristics, whose appearances had been separated in time by the same number of years or multiples thereof; and he found that the comets of 1531, 1607, and 1682 seemed to fit the bill. In 1695, he told Newton that he thought these were reappearances of the same comet.

However, the intervals separating them, though similar, were not identical. Halley realized that this was because the orbit would have been modified, whenever the comet passed near a major planet during its passage through the solar system and in so doing experienced the planet's gravitational pull; and he predicted the same comet's return 'about the end of the year 1758 or the beginning of the next'.

Could it be that these portentous apparitions were, after all, as lawlike as planets? In the summer of 1757, Alexis-Claude Clairaut (1713–65) and two associates laboured against the clock to calculate in greater detail how the orbit of the comet would have been modified in 1682 as it passed close to Jupiter when departing the solar system; and finally they were able to predict that the returning comet would swing around the Sun within a very few weeks of mid-April 1759.

A newly arriving comet was indeed seen on Christmas Day of 1758, and it rounded the Sun on 13 March 1759. Crucially, the characteristics of its orbit were closely similar to those of the three comets Halley had studied: all four comets were one and the same. To the astonishment of astronomers and public alike, Newtonian mechanics had predicted the return of 'Halley's Comet' after an interval of three-quarters of a century.

Meanwhile, much mathematical effort was being expended on analysing the complex behaviour of the Moon. This was motivated in part by mathematical curiosity, but there was also a much more serious purpose. The lives of sailors at sea depended on their knowing where they were, especially at night. To determine the ship's latitude was relatively straightforward: the navigator measured the altitude of the north celestial pole at night (or, less directly, the altitude of the Sun at midday). Longitude – the time-change all too familiar to air travellers today – was much more difficult, for how was one to compare local time with a standard time (today, Greenwich Mean Time)? By the early 18th century, pendulum clocks were performing acceptably well on land, but they were useless at sea.

From time to time down the centuries, attempts had been made to exploit a suggestion of Hipparchus, that the difference in longitude between cities could be determined by comparing the local times of an eclipse of the Moon, viewed simultaneously from the two locations; but such eclipses were too rare to be of use to navigators. Galileo had proposed instead the eclipses of Jupiter's moons, which are much more common; and later in the 17th century accurate tables of the Jovian satellites allowed this method to be used successfully on land. But such eclipses – still rare enough in all conscience – were well-nigh impossible to observe from on board ship.

Other methods were tried that varied from the near-hopeless to the downright bizarre, and eventually the serious alternatives reduced to two: the development of a chronometer that would keep accurate time at sea, and the use of the Moon's rapid movement against the background stars as the analogue of the movement of a clock's hour-hand against the hour-numerals of the dial. The prize offered by the British Parliament for a practical solution to the problem of longitude at sea would make the recipient rich beyond the dreams of avarice.

The chronometer was work for clockmakers, men who worked with their hands, chief among them John Harrison (1693–1776). Meanwhile, university-trained astronomers and mathematicians struggled to perfect the method of 'lunar distances'. To implement this method, the navigator would first have to determine the Moon's current position in the sky – in practice, its position relative to nearby stars. For this he needed an accurate star catalogue, and an accurate instrument with which to measure the angles between the Moon and convenient stars. He then required reliable lunar tables that would convert this observed position of the Moon into standard time, which he could compare with his local time to give him his longitude. Errors in star positions, in the measurement of the angles, and in the lunar tables would all increase the distance between where the ship in fact was and where the navigator reckoned it to be, and so it was crucial that each of the three be reduced as far as was humanly possible.

The Royal Observatory at Greenwich was founded in 1675 expressly to meet navigators' need for an accurate star catalogue; and the posthumous publication in 1725 of Flamsteed's 'British Catalogue' of 3,000 stars, which improved on Tycho's naked-eye star catalogue by a whole order of magnitude, was the first Astronomer Royal's fulfilment of this need. The requirement for an accurate instrument for measuring angles, suitable for use at sea, was solved by the invention in 1731 of a double-reflection quadrant, ancestor to the sextant. It was now up to mathematicians – all of them, as it happened, French or German – to perfect Newton's lunar theory so that accurate tables of the Moon's position could be calculated months in advance and supplied to navigators. Eventually the Göttingen professor Tobias Mayer (1723–62) developed tables good enough to earn his widow £3,000 of the prize on offer in Britain, and these allowed the then Astronomer Royal, Nevil Maskelyne (1732–1811), to publish in 1766 the first of the annual volumes of *The Nautical Almanac*.

Meanwhile, however, Harrison was producing a succession of masterly chronometers, the first of which was taken for trial to Lisbon and back in 1736. The results were encouraging and Harrison was awarded £250 to fund further research and development. Things continued in this way for nearly 30 years, until in 1764 Harrison sailed with his fourth chronometer to Barbados and back, after which he was awarded half of the £20,000 originally offered as prize money. As soon as suitable chronometers could be constructed in quantity, they became the preferred solution to the problem of longitude, and astronomers found themselves with a new role, manning observatories at major ports and dropping a time-ball at noon (or 1 p.m.) so that navigators could check their chronometers before setting sail. Harrison's own chronometers – poetry in motion – may be seen in operation today in the National Maritime Museum at Greenwich.

Newton had seen the curiously large gap separating the small inner planets from the massive Jupiter and Saturn as evidence of Providence's concern to preserve the solar system from disruption, but Kepler had earlier toyed with the idea that the gap was occupied by a planet as yet undiscovered. By the 18th century, references to 'the known planets' (with the implication that there might be others still unknown) were not uncommon, and speculations about a possible 'missing' planet were fuelled by the discovery of a curious arithmetical pattern in the distances of the planets from the Sun. In his *Elementa astronomiae* of 1702, the Oxford professor David Gregory (1659–1708) had put these distances as proportional to 4, 7, 10, 15, 52, and 95; and by slightly modifying two of the numbers, Johann Daniel Titius (1729–96) of Wittenberg made them equal to 4, 4 + 3, 4 + 6, 4 + 12, 4 + 48, and 4 + 96. These numbers have the form $(4 + 3 \times 2^n)$. The proposed pattern was enthusiastically adopted by the young German astronomer Johann Elert Bode (1747–1826), and it is today spoken of as Bode's Law. Titius and Bode agreed that there must be, or have been, a body or bodies corresponding to the term $4 + 3 \times 2^3$.

In 1781 came a wholly unexpected development: William Herschel (1738–1822), an amateur observer of whom we shall have much to say, was familiarizing himself with the brighter stars when he came across a 'curious' object that turned out to be the planet we now know as Uranus. When mathematicians were able to determine its orbit, they made the remarkable discovery that its distance from the Sun matched the next term in the sequence, $4 + 3 \times 2^6$. This was enough to convince the court astronomer at Gotha, Baron Franz Xaver von Zach (1754–1832), of the validity of the pattern, and he began to search for a planet corresponding to the term $4 + 3 \times 2^3$. Having no success, in 1800 he held a meeting with a group of

friends to discuss how best to proceed, and they divided the zodiac –
the region where any planet was likely to be – into 24 zones; each
zone was to be assigned to a particular observer whose duty
would be to police his zone and look out for any 'star' of no fixed
abode.

One of the intended patrolmen was Giuseppe Piazzi (1746–1826)
of Palermo Observatory in Sicily. Piazzi was currently at work on a
star catalogue, and his careful method of working required him to
remeasure the position of each star on a subsequent night. On
1 January 1801, before the invitation from von Zach and company
reached him, Piazzi measured the position of an eighth-magnitude
'star'; and when he came to remeasure it, he found that it had
moved.

Piazzi was able to track the object for only a few weeks before he
lost it in the glare of the Sun, and it was thanks to the emerging
mathematical talent of Carl Friedrich Gauss (1777–1855) that it was
recovered, by von Zach, at the end of the year. Ceres, as Piazzi
called it, matched the missing term in Bode's Law, but it was tiny:
Herschel (rightly) thought it smaller even than the Moon. Worse
still, three more such objects, also tiny, and also matching the
missing term, were discovered in the next few years. Herschel
proposed calling the members of this new species of heavenly
body 'asteroids'. Wilhelm Olbers (1758–1840), a physician and
astronomer also involved in the search for the missing planet,
thought they might be fragments of what had once been a full-sized
planet.

The search continued for a number of years, but it proved fruitless,
and was eventually abandoned. Not until 1845 was another
asteroid discovered, by K. L. Hencke, a German ex-postman, and
his second success two years later revived interest. By 1891 more
than 300 asteroids had been found, and photography was now
simplifying the search. Max Wolf (1863–1932) at Heidelberg would
photograph a large star field over several hours with a telescope

that tracked the rotation of the sky; stars would appear as points of light, but an asteroid would leave a trail as it moved relative to the stars.

Had Olbers been correct, the orbit of each asteroid would – initially at least – have passed through the place where the planet disintegrated, and also through the matching place on the opposite side of the Sun. This proved not to be the case, and astronomers now think that asteroids, whose combined mass is only a fraction of that of the Moon, are objects that failed to coalesce into a planet because of the attractive pull of Jupiter.

The discovery of Uranus had extended the sequence of Bode's Law, but the planet's movements quickly proved puzzling. Early determinations of its orbit were greatly simplified by the discovery that it had been observed (and listed as a star) as long ago as 1690; yet the planet soon began to deviate from its predicted path. Various explanations were proposed, and these were eventually narrowed down to two – either the formulation of the inverse-square law required amendment at such distances, or Uranus was being pulled by a planet as yet undiscovered – and then to one: the undiscovered planet. By the 1840s two talented mathematicians were at work at their desks, putting pen to paper and hoping, by mere calculation, to tell astronomers where to look for this unknown (and previously unsuspected) satellite of the Sun.

The younger of the two was John Couch Adams (1819–92), a graduate of Cambridge, where James Challis (1803–82) was professor. In the autumn of 1845, at Challis's suggestion, Adams visited Greenwich to explain his calculations to the Astronomer Royal, George Biddell Airy (1801–92). By an unlucky chance he failed to see Airy in person, but he left a summary of his results. Next summer Airy was astonished to receive from Paris a copy of a paper by Urbain Jean Joseph Le Verrier (1811–77) predicting the presence of a planet in almost the same position. Airy was of the view that research was not the purpose of the national observatory that he directed, but he asked Challis to institute a search in Cambridge.

The only procedure open to Challis was to plot the star-like objects in a region of sky with care, and then to return to the same region at a later date to see if one of them had moved. This was inevitably a tedious and time-consuming process, and Challis saw no urgency. Unfortunately for the future of Anglo-French relations, Le Verrier had asked astronomers at the Berlin Observatory to make a search, and they – unlike Challis – had a copy of the relevant sheet of the Berlin Academy's new star atlas. They were therefore in a position to compare the stars in the sky with those in the atlas, and within minutes of starting their search on 23 September 1846 they found a star-like object not on the sheet. It was the missing planet.

Challis, it later transpired, had noted the same 'star', but he had not yet returned to the same region of sky to remeasure its position. To the English, Adams's moral claim to the discovery of the planet Neptune was equal to that of Le Verrier, but that was not how the French saw it. But whatever the correct view of this priority dispute, the triumph of Newtonian mechanics was now complete.

This pleasing state of affairs was not to last. Like Uranus, Mercury had an unexplained feature to its orbit: its point of nearest approach to the Sun was advancing in longitude more rapidly than expected, by about one degree per century – a tiny amount, but one that nevertheless called for explanation. Le Verrier, not surprisingly, suspected yet another unseen planet; and in September 1859, he announced that a planet the same size as Mercury but at half the distance from the Sun (and therefore difficult to observe) would be one possible explanation of the phenomenon. It happened that an unknown French physician named Lescarbault had earlier that year seen an object crossing the Sun (or so he thought), and when he read of Le Verrier's prediction, he wrote to him. Le Verrier satisfied himself as to Lescarbault's reliability, and named the planet that the physician had supposedly seen Vulcan. Various alleged sightings of Vulcan followed, but few were convincing; and by the end of the century Vulcan had been rejected as spurious. In 1915 Albert Einstein was to show that the anomalous behaviour of Mercury was implied by his General Theory of Relativity: there was more to the universe than was dreamt of in Newtonian philosophy.

# Chapter 6
# Exploring the universe of stars

Until 1572, astronomers viewed the 'fixed' stars – fixed and unchanging, that is, in position but also in brightness – as little more than a backdrop to the motions of the planets. In fact, of course, stars do have individual, or 'proper', motions across the sky, but the scale of interstellar distances is so immense that light from even the nearest stars takes years to reach us. As a result, proper motions are almost imperceptible, except over very long time-scales; and so the Renaissance observer saw the stars in positions seemingly no different (except for the overall effect of precession) from those assigned them by Ptolemy.

That no changes in brightness had been noticed is perhaps more surprising. Although most stars, like the Sun, are almost invariable, a minority appear brighter at some times than at others. Some diminish in brightness as they are eclipsed by a companion star, while others are subject to major physical changes, whether regular or irregular. But it so happens that none of these variable stars was bright enough for the variations to force themselves on the attention of an Aristotelian observer of the Middle Ages, convinced as he was that the celestial regions were immutable. Why look for change when you already know that change is impossible?

The result was that it took a wake-up call from Nature – the appearance of the new star, or nova, of 1572 (as seen by Tycho

Brahe, see Chapter 4) – to draw attention to the stars as objects that change and are therefore of interest. Another such nova shone forth in 1604, and this one generated alarm and despondency across Europe. For the first time in eight centuries, the slow-moving planets Jupiter and Saturn were in conjunction in the fateful 'fiery trigon' of the zodiac; and no sooner had they been joined there by Mars, than this new star blazed forth in their midst – the most ominous astrological event imaginable.

No one could now doubt that changes occurred in the heavens. Indeed, there was talk of another nova that had appeared in the constellation of the Whale, but this was fainter, and had been seen by only a single observer before it faded and vanished. In 1638 the Whale was host to a second nova (or so it seemed); like its predecessor, it faded and vanished – but before its discoverer could publish his account of it, it astonished him by reappearing. It continued to vanish and reappear at intervals, and in 1667 Ismael Boulliau (1605–94) announced that this 'wonderful star' was reaching maximum brightness every 11 months: its behaviour was to some extent predictable, and therefore lawlike.

Boulliau went on to offer a physical explanation of variable stars that was ingenious, indeed too much so. He pointed out that variations in sunspots show that the Sun itself – by now recognized as simply the star nearest to us – is, strictly speaking, variable. Furthermore, the rotation of sunspots had demonstrated that the Sun as a whole rotates; and no doubt other stars did likewise. Imagine, then, a rotating star with extensive dark patches instead of mere spots; whenever a dark patch is facing towards us, we shall see the star diminished in brightness, and this will happen regularly, with every rotation of the star. But if the patches themselves vary irregularly, as do sunspots, then this will result in irregular changes in brightness. In this way Boulliau was able to explain both regular and irregular variations. Indeed, so successful was he that the physical explanation of variable stars ceased to be a problem, and astronomers contented themselves with announcing their discovery of variations in particular stars. But these claims could not easily be verified or falsified; the number of allegedly variable stars grew by leaps and bounds, and the whole subject fell into something of disrepute.

Towards the close of the 18th century, the task of verifying such claims was simplified when William Herschel published a series of 'Catalogues of the Comparative Brightness of the Stars'. In these lists Herschel carefully compared stars with neighbouring ones of similar brightness, so that a variation in one of the stars would reveal itself by disturbing the published comparisons. Herschel was generalizing a method involving sequences of stars arranged in order of brightness, which had been developed in the early 1780s by two amateurs who were neighbours in the city of York in the north of England. Edward Pigott (1753–1825) was the son of an accomplished observer; his youthful friend, John Goodricke (1764–86), was a deaf-mute who eagerly accepted the invitation to collaborate in the study of variables.

One of the stars they scrutinized was Algol, which a century earlier had twice been reported as fourth magnitude instead of the usual second. On 7 November 1782 it was second magnitude as usual, but five nights later it was down to fourth; the next night it was back to second. Changes at this rate were unprecedented, so both men now kept watch on the star. On 28 December their efforts were rewarded when they saw Algol start the evening at third or fourth magnitude, and brighten to second before their very eyes. Pigott instantly suspected that Algol was being eclipsed by a satellite, and next day sent Goodricke a note in which he calculated the future orbit of the hypothetical satellite, on the assumption that in the 46 days between 12 November and 28 December, the satellite had completed either one or two orbits. In fact, their observations during the coming months showed that the satellite – if indeed this was the explanation – orbited Algol in less than three days, a phenomenon hitherto unknown to astronomy.

Pigott generously left it to his handicapped friend to make the formal announcement to the Royal Society, but Goodricke, who was still in his teens, merely mentioned the eclipse theory as one possible cause, alongside the traditional dark patches. Pigott had in fact been right – Algol is indeed eclipsed by a companion – but the two friends eventually reverted to the dark patches explanation, perhaps because they wrongly thought they had evidence of irregularities in the light curve of Algol, or because the three other short-period variables that they discovered could not be explained by the eclipse theory. Two, in fact, were cepheids, pulsating stars that climb rapidly to a maximum brightness and then slowly decline, and which would one day be employed as distance indicators by Edwin Hubble and his contemporaries.

The outcome was that 18th-century astronomy yielded a new class of variable stars, those with periods of only a few days, but little progress in understanding the underlying physical causes of these phenomena.

Pigott and Goodricke had seen Algol vary in brightness in a matter of hours. By contrast, changes in position are detectable only in the very long term. Relatively few stars have a proper motion of as much as one second of arc per annum, and the largest known is only just over ten seconds of arc. Such motions can be detected only by comparing the modern position of the star with its position as recorded in a catalogue from an earlier epoch; and, other things being equal, the longer the interval of time since the earlier catalogue, the more accurate will be the resulting value for the proper motion. Unfortunately, however, things are not equal: standards of accuracy decline as we travel backwards in time, and any uncertainty in the position of the star as stated in the earlier catalogue will affect the precision with which its proper motion can be determined.

The only star catalogue from antiquity is that in Ptolemy's *Almagest*; and it was while using this catalogue in 1718 to determine the rate of change of the obliquity of the ecliptic – the angle at which the ecliptic is inclined to the celestial equator – that Edmond Halley realized that three stars must have moved independently of the rest.

As it happened, it would not have been easy for Halley to pursue the question further, for the only other past catalogue of value was Tycho's. This was much more accurate than Ptolemy's; but it was little more than a century old, and its author had dealt crudely with refraction, the bending of starlight as it enters the Earth's atmosphere (which affects the observed position of the star in the sky). Future generations, however, would be able to take the 'British Catalogue', compiled with the greatest care by John Flamsteed at Greenwich, as their starting point in time from which to measure proper motions.

Or so it seemed. But then, in 1728, James Bradley (1693–1762) announced a wholly unexpected complication: the 'aberration of light'. The speed of light is very great, but it is nevertheless finite, as had been shown late the previous century by observation of eclipses of the satellites of Jupiter. These seemed to occur ahead of schedule when the planet was near Earth and the light bearing news of the eclipse had less far to travel, and behind schedule when Jupiter was away on the far side of the Sun.

By comparison, the speed of the Earth in its orbit around the Sun is small, but it is nevertheless big enough to affect the observed positions of the stars. A star appears to an observer to lie in the direction from which its starlight arrives; and this direction alters (slightly) with changes in the direction in which the Earth is moving – just as rain that is in fact falling vertically seems to strike our faces from the direction towards which we are currently moving.

How Bradley came to discover aberration we shall see later in this chapter. The implication of his discovery was that even the British

Catalogue would be seriously defective as the starting point in time for the measurement of proper motions. A further defect became apparent in 1748, when Bradley announced that the Earth's axis 'nutated', or nodded. This was a result of the varying gravitational pulls of the Sun and Moon on an Earth that is not a perfect sphere, and it led to a movement of the very coordinate system used in the measurement of stellar positions.

Bradley himself became Astronomer Royal in 1742, and from 1750, until his health began to fail, he carried out a programme of observations in which he meticulously recorded all the circumstances that might influence the observed position of a star. But he himself did not live to 'reduce' his observations – to make the calculations necessary to derive the true positions of his stars. The reductions had to wait until 1818, when the great German mathematician Friedrich Wilhelm Bessel (1784–1846) published the aptly named *Fundamenta astronomiae*, with over 3,000 stellar positions for 1755, a convenient date midway through Bradley's observing campaign. From then onwards, 19th-century astronomers were able to compare the current position of a star with its 1755 position as given in the *Fundamenta* and so determine how far across the sky the star had moved each year in the interval.

Bradley himself pointed out in 1748 that all proper motions are relative: we do not observe how a star is moving in absolute space, but how it is moving relative to us. Twelve years later, Tobias Mayer discussed the implications of this. If every star except the Sun were at rest, then the motion of the solar system through space would reveal itself to us as a pattern of (apparent) movements among the stars. Therefore, any pattern among the known proper motions is likely to reflect a motion of the solar system; the residual motions will be those of the individual stars themselves.

A modern analogy illustrates the form such a pattern might take. If we drive a car in a city at night, a cluster of distant traffic lights appear bunched together, but they seem to move apart as we approach. At the same time, the street lights on our left appear to move anti-clockwise, while those on our right seem to move clockwise.

Mayer himself could find no such pattern in the (unreliable) proper motions known to him, but in 1783 William Herschel – for once working entirely at his desk – believed he had found a pattern that implied that the solar system is moving towards the constellation Hercules. Today there is no doubt that his conclusion was correct, but his argument does not withstand close scrutiny. A generation later, Bessel had unique access to reliable proper motions in the months during which his *Fundamenta* was in press, and he possessed all the mathematical talent necessary to unravel any pattern; yet he drew a blank.

It was only in 1837 that astronomers became convinced that a solution was in sight. In that year, F. W. A. Argelander (1799–1875), professor of astronomy at Bonn, published an analysis of no fewer than 390 proper motions which he divided by size into three groups, and each group independently yielded a direction not far from that proposed by Herschel.

His conclusions were quickly confirmed by analyses by other astronomers; yet these all depended on the same basic data – Bradley's observations of the stars visible from England. However, Nicholas-Louis de Lacaille (1713–62) had visited the Cape of Good Hope in 1751–53 and determined the positions of nearly 10,000 stars, and 19th-century positions for some of these southern stars were now becoming available. In 1847, 81 resulting proper motions – data that owed nothing whatever to Bradley – were analysed by an actuary, Thomas Galloway (1796–1851), and he derived a direction similar to that based on data for northern stars. Since then there has been no doubt that the solar system is moving in the direction of Hercules, and subsequent analyses of the known proper motions – a list that has increased greatly in number and accuracy with the passage of time – have served only to refine this conclusion.

How far away are the stars? For Ptolemy in antiquity, and for Tycho Brahe in the late 16th century, the fixed stars were just beyond the outermost planet. But if Copernicus was right, then every six months we observe the stars from opposite ends of a huge baseline whose length is twice the radius of the Earth's orbit around the Sun (two 'astronomical units'). As we have seen, even Tycho with his precision instruments was unable to detect the apparent movement among the stars that would result ('annual parallax'), and he very reasonably saw this as refutation of the heliocentric hypothesis.

Part of the problem lay in the nature of the observations: as the months pass, seasonal changes in temperature and humidity will cause an instrument to warp, refraction will vary in response to changes in air pressure, and so on. Galileo, ingenious as ever, saw a way to overcome these difficulties. Suppose two stars lie in almost the same direction from Earth, and suppose one is much further than the other.

The further will have a much smaller parallax than the nearer; this means that we shall not go far wrong if we ignore the parallax of the further altogether, and take it as a quasi-fixed point in the sky from which to measure the parallax of the nearer star. The advantages of this will be great, for the two stars will be equally affected by any warping of the instrument, changes in refraction, and so forth, and the effects of such complications will be eliminated from consideration.

Galileo, indolent as ever, made no attempt to follow up his own suggestion, and it would be many years before it bore fruit. Meanwhile, René Descartes convinced the learned world that stars are suns and the Sun merely our local star, and this suggested a different approach to the problem of stellar distances.

If space is perfectly transparent, light falls off with the square of the distance. Therefore, if the Sun is removed to a thousand times

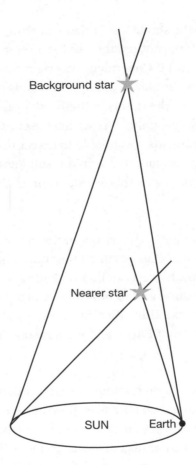

**15. Galileo's method of detecting annual parallax, by measuring the apparent annual motion of a nearby star relative to a background star**

further than it is at present, it will appear to shine with one-millionth of its present brightness. Suppose now that the stars are not merely of similar nature to the Sun but physically identical to it, so that Sirius (for example) is the physical twin of the Sun. Then if Sirius is found to be one-millionth the brightness of the Sun, and provided space is transparent, we shall know that Sirius is one thousand times further away than the Sun.

But how is one to make the comparison between the glare of the bright Sun and the faint light of a star? The Dutch physicist

Christiaan Huygens (1629–95) put a screen between himself and the Sun, and made a tiny hole in it. His intention was to vary the size of the hole until the portion of the Sun visible through it was equal in brightness to Sirius, and then to calculate what fraction of the Sun was thus visible. It was a crude method, but his result – that Sirius is 27,664 astronomical units from us – was the only estimate available in print for over a quarter of a century after its publication in 1698, and so was widely quoted. Evidently the stars were a long way away.

Meanwhile, unknown to all but a tiny circle of intimates, Isaac Newton had made much better progress, by using an ingenious proposal of the Scottish mathematician James Gregory (1638–75). In a little-noticed book published in 1668, Gregory proposed simplifying the photometric comparison by using a planet as a substitute for Sirius. One was to wait until the planet was equal in brightness to Sirius, and then use one's knowledge of dimensions within the solar system to compare the Sun's light that comes to us directly with the Sun's light that comes to us via the planet. Working on these lines, Newton put Sirius at one million astronomical units. As it happens, Sirius is rather more than half that distance, and so among Newton's intimates, the enormity of the distances that separate the Sun from the nearer stars was now fully appreciated.

But such estimates, based on the provisional assumption that every star is a twin of the Sun, were no substitutes for the actual measurement of the annual parallax (and hence distance) of specific stars. It occurred to Robert Hooke that since the star Gamma Draconis passed directly overhead his lodgings in London, its light would then be unaffected by atmospheric refraction. He tried to avoid the danger of seasonal warping of the observing instrument by incorporating the components of a telescope into the actual fabric of his house. Although telescopic astronomy was in its infancy, Hooke designed and built a telescope to observe one single star, at just one moment during its passage, and for one purpose only.

Hooke's ingenuity was not matched by his perseverance: he made a mere four observations, in 1669, before illness and an accident to the telescope lens brought his efforts to an end. But his method had much to commend it, and in the mid-1720s a prosperous English amateur, Samuel Molyneux (1689–1728), decided to make another attempt to measure the annual parallax of Gamma Draconis. He invited James Bradley to join him, and he commissioned a special 'zenith sector' from a leading maker, George Graham. The sector with its vertical telescope was mounted on a chimney stack in Molyneux's house, and when a star passed overhead the tube of the telescope was tilted slightly to bring the star into the middle of the field of view; the angle of tilt was measured against the scale of the sector to give the angular distance of the star from the vertical.

A simple calculation showed that Gamma Draconis should reach an extreme southerly position a week before Christmas, and so it was a surprised Bradley who on 21 December saw it pass overhead markedly further south than it had a week earlier. By March, when it should have been moving north, it was some 20 seconds of arc south of its December position. The star then stopped and back-tracked, passing through its December position in June, and reaching an extreme northerly position in September.

The two friends debated various explanations – was there a movement in the axis of the Earth and therefore in the coordinate system by which the star's position was being measured, or was the atmosphere of the Earth distorted by the planet's passage through space so that atmospheric refraction was unexpectedly affecting the measures? – but without success. Bradley decided to commission from Graham another zenith sector, this time with a wider field of view that would bring more stars under scrutiny, and with it he established the patterns of the stellar movements; but their explanation eluded him. Then, one day, while on a boat on the Thames, he noticed that the weather vane changed direction as the boat put about – not of course because the wind had veered, but because the boat had altered course. The starlight, he now realized, was likewise reaching the observer from changing directions, because the observer was altering course as the Earth orbited the Sun.

This discovery of aberration, announced to the Royal Society in 1729, was momentous, for several reasons. It was the first direct proof of the Earth's motion around the Sun. Because all stars were similarly affected, it showed that the velocity of light was a constant of nature. It revealed (as we saw earlier) a wholly unexpected error in past measurements of stellar positions, Flamsteed's among them. And because even Bradley's zenith sector, for all its accuracy, was unable to detect annual parallax, the stars must lie at distances of at least 400,000 astronomical units.

Only the previous year, the posthumous publication of Newton's *The System of the World* had made public his estimate – based, it is true, on the working hypothesis of physical uniformity among the stars – that Sirius lay at one million astronomical units. These two results – the one giving an actual distance but on the basis of a questionable hypothesis, the other a minimum distance based on direct measures – combined to convince astronomers that the scale of stellar distances was at last understood.

The unwelcome implication was that annual parallax must be at most a second or two of arc, an angle so tiny as to represent the width of a coin at a distance of kilometres. Such a minute movement, taking place over a period of months, would be well-nigh impossible to detect, and the next generation of astronomers displayed little enthusiasm for such a hopeless task. William Herschel in the 1770s and 1780s collected a great number of double stars, ostensibly for use in Galileo's method of measuring parallax; but he was playing the natural historian, collecting specimens that others might one day put to use. Most astronomers preferred to spend their time on more promising lines of enquiry.

In any case, John Michell (*c*.1724–93) had shown in 1767 – unknown to Herschel – that the number of double stars was so large that most must be true companions in space (binary stars), lying at the same distance from the observer, and therefore useless for Galileo's method of measuring parallax. Herschel himself was to confirm Michell's claim when he re-examined some of his doubles around the turn of the century, and found instances where the two stars had orbited relative to each other. A generation later William's son John was to be one of those who confirmed that their orbits were indeed ellipses, and that the force binding the companion stars together was therefore Newtonian attraction. Newton had claimed that attraction was a universal law, but this was the first evidence that the law applied outside the solar system.

Meanwhile, astronomers found themselves in a situation where, as telescopes improved, the two coordinates of a star's position on the heavenly sphere were being measured with ever increasing accuracy, whereas little was known of the star's third coordinate, distance, except that its scale was enormous. Even the assumption that the nearest stars were the brightest was being called into question, as the number of known proper motions increased and it emerged that not all the fastest-moving stars were bright.

An extreme example of this was found early in the 19th century when first Piazzi and then Bessel found that the relatively faint star 61 Cygni was travelling across the sky with the exceptional speed of over 5 seconds of arc per annum. Surely this showed that the star must be near, despite its modest brightness?

Annual parallax is of course inversely proportional to distance, and it was crucial that observers attempting to measure this parallax concentrate their efforts on the stars closest to Earth. In 1837, after several claims to success in measurement had proved ill-founded, the German-born Wilhelm Struve (1793–1864) proposed three criteria of nearness: was the star bright, was its proper motion sizeable, and – if it happened to be a binary star – did the two components appear widely separated in relation to the time they took to orbit each other?

At his observatory at Dorpat (now Tartu in Estonia) Struve was privileged to possess a magnificent refracting telescope by Joseph Fraunhofer (1787–1826). Its object glass was no less than 24 cm in diameter and of exceptional quality, and its mounting was 'equatorial', its axis pointing to the north celestial pole so that the observer needed to rotate only this one axis to keep the telescope aligned on a star. In 1835 Struve had selected for his parallax measurements the star Vega, which is very bright and has a large proper motion, and in 1837 he announced the results of 17 observations, from which he inferred a parallax of one-eighth of a second of arc. Three years later he reported on 100 observations, and this time inferred a parallax of one-quarter of a second. But given the long history of spurious claims, going back to Hooke, astronomers had yet to be convinced.

Meanwhile, Bessel, at Königsberg, was equally fortunate in his instrumentation. His Fraunhofer refractor was not as large, the objective having a diameter of just 16 cm. But its maker, not content with achieving a lens of high quality, had then taken his courage in both hands and cut it into two semicircular pieces of glass that could move against each other along their common diameter. Each half showed a complete image, but of only half the brightness. If the telescope was turned towards a double star, the pair of stars would appear in both halves, and the observer could then slide the halves relative to each other until one star in one image coincided with the other star in the other image. The amount of displacement necessary was then a very accurate indication of the angle separating the two stars. Because such instruments were often used to monitor changes in the apparent diameter of the Sun, they were known as heliometers.

Bessel selected for scrutiny 61 Cygni, known as the 'flying star' because of its large proper motion. In 1837 he subjected the star to an unprecedented examination, observing it for over a year as many as 16 times in a single night – even more often if the 'seeing' was particularly good. The following year he was able to announce that it had a parallax of about one-third of a second of arc. What carried conviction was the way in which the plotted graph of his observations matched the expected theoretical curve. It was, John Herschel told the Royal Astronomical Society, 'the greatest and most glorious triumph which practical astronomy has ever witnessed'. The stellar universe now had a third dimension; and the number of successful measurements of annual parallax would multiply in the decades to come.

But what was the large-scale structure of this universe? Newton's *Principia* had almost nothing to say about the stars, and it seems that he had given questions of cosmology little thought prior to 1692, when he received a letter from a young theologian, Richard Bentley (1662–1742). Bentley had given a series of lecture/sermons on science and religion, and before putting these into print he wanted to know the views of the author of that densely mathematical book that everyone respected but no one understood. Bentley had no time for the Cartesian position whereby God had created the universe and left it to run itself; but he wanted to know what could be said in support of this, and so he asked Newton what would happen in a universe in which the matter was initially distributed with perfect symmetry. Newton, not realizing that Bentley intended the symmetry to be literally perfect, replied that wherever the matter was more concentrated than usual, its gravitational pull would attract the surrounding matter and so lead to still greater concentration. Corrected, and irritated, he conceded that in a *perfectly* symmetric universe the matter would have no reason to move one way rather than another; but he commented that a perfect symmetry was as plausible as having infinitely many needles all standing on their points on an infinite mirror. 'Is it not as hard', retorted Bentley, 'that infinite such Masses in an infinite space should maintain an equilibrium?' In other words, how was it that the stars were all 'fixed' and motionless, when each was supposedly pulled by the gravitational attraction of all the others?

Newton was now squarely confronted with the paradox underlying the claim in the *Principia* that attraction is a universal law of nature; for it seemed that, even after observations of centuries, the stars were as fixed as ever. Curiously, Newton was (as we have seen) the only person to have a correct appreciation of the scale of interstellar distances; yet it did not occur to him that, because the stars were so enormously far away, any motions they had would be well-nigh imperceptible. Instead, he continued to believe that the stars – the *fixae* – were motionless, and his problem was to explain how this could be so.

His solution to the paradox is to be found in drafts for an intended second edition of the *Principia* that he abandoned when he left Cambridge for a post in London. We remember that he saw the finite system of planets orbiting the Sun as having been planned by Providence to provide a stable environment for mankind, though the stability was not perfect and therefore Providence would eventually have to intervene, to prevent gravity from undermining the system. The system of the stars was likewise stable; but this, he argued, was because the stars were infinite in number and their distribution (almost) symmetric: each star was initially at rest, and it remained so because it was pulled equally in every direction by the other stars.

Yet a glance at the night sky would show that the symmetry was not in fact perfect; and indeed Newton needed ingenuity to provide evidence for at least the semblance of symmetry among the nearer stars. But he did not see imperfect symmetry as a problem: instead, it was another area for regular interventions by Providence, as a result of which the stars would be restored to their earlier order.

Newton had focused on the dynamics of the stellar universe, but what of the light sent to us by these same stars? This question was put to Newton around 1720 by a young physician of his acquaintance, William Stukeley (1687–1765). Galileo's telescope, a century before, had confirmed that the Milky Way resulted from the combined light of innumerable tiny stars; but – curiously – there had been little subsequent interest in the three-dimensional distribution of stars that gives rise to this phenomenon, and it did not occur to Newton that the Milky Way was disproof of his claim that the universe of stars is symmetric.

Stukeley, however, speculated that the stars bright enough to be individually visible might together form a spherical assemblage, while the stars of the Milky Way formed a flattened ring surrounding this sphere – in effect, a stellar analogue to Saturn and its ring. In response Newton hinted that an infinite symmetric universe of stars was to be preferred; to which Stukeley – not knowing that Newton was secretly committed to this very concept – retorted that in such a universe 'the whole hemisphere [of the sky] would have had the appearance of that luminous gloom of the milky way'.

Early in 1721, Stukeley and Halley took breakfast with Newton, and they discussed questions of astronomy. These must surely have included the possibility of an infinite universe of stars, for a few days later Halley read to the Royal Society the first of two papers on the subject. When his papers were published in *Philosophical Transactions*, Newton's model of the universe came at last – if anonymously – into the public domain.

In one of the papers Halley remarked carefully: 'Another Argument I have heard urged, that if the number of Fixt Stars were more than finite, the whole superficies of their apparent Sphere would be luminous.' He had his own solution to Stukeley's misgivings, but it was flawed. It was not until 1744 that a correct analysis of light in an infinite and nearly symmetric universe was published. The Swiss astronomer J.-P. L. de Chéseaux (1718–51), pointed out that at the distance of the nearest stars there was (so to speak) room for a given number of stars such that no two were unduly close together; and that in aggregate these stars filled a certain (tiny) area of celestial sphere. At twice the distance there was room for four times as many stars, but each would be one-quarter the brightness and one-quarter the apparent size. In aggregate, therefore, they would fill the same area of the sky as before, and to the same level of brightness. At three times the distance, the stars would fill a further area of sky with light; and similarly for each succeeding step, until eventually the entire sky was ablaze with light.

Or so one might think (and indeed modern astronomers see the darkness of the night sky as posing 'Olbers's Paradox'). But Chéseaux pointed out – as did Olbers in 1823 – that this reasoning assumes that all the light that sets out from a given star reaches its destination; whereas even a minute loss of light, repeated at each stage of the journey, would effectively reduce the very distant stars to invisibility. For neither Chéseaux nor Olbers was there any paradox.

Nor was there for later 19th-century astronomers, even though by now it was realized that an interstellar medium that intercepted light would itself heat up and begin to radiate. There were plenty of other ways out of the difficulty, such as the existence of etherless vacua across which no light could pass. Only in our own time has the darkness of the night sky been elevated to the status of a paradox, and those who christened it were unaware that the question goes back beyond Olbers, to Chéseaux, Halley, and ultimately to the physician Stukeley.

Meanwhile, amateur speculators began to puzzle over the Milky Way. In 1734 Thomas Wright of Durham (1711–86) gave a public lecture/sermon in which he presented his highly personal cosmology. The Sun and the other stars, he told his audience, orbited around the Divine Centre of the universe, occupying as they did so a spherical shell of space outside of which was the Outer Darkness; and he concentrated the minds of his audience by

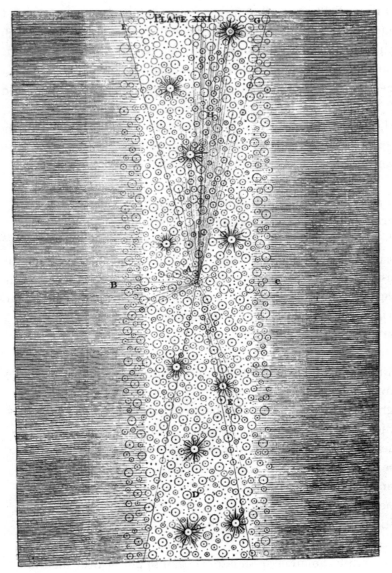

16. A sketch used by Wright to help readers understand his preferred model of our system of stars. In this imaginary universe there is a layer of stars bounded by two parallel planes. An observer at *A* would see only a handful of near (and therefore bright) stars when looking outwards from the layer, in the direction of *B* or *C*; but when looking along the layer, in the directions of *D*, *E*, etc., the observer would see innumerable stars, near and far, whose light would merge to give a milky effect. From Thomas Wright, *An Original Theory of the Universe* (1750).

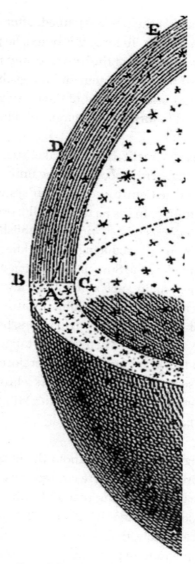

17. Wright's preferred model of the star system to which the solar system belongs. The space occupied by the stars has the form of a spherical shell, whose radius is so vast that its curvature is imperceptible to the human observer located at *A*. For the observer, therefore, the inner and outer surfaces of the layer of visible stars approximate to parallel planes. As before, the observer sees only a handful of near and therefore bright stars when looking in the direction of *B* or *C*, but sees innumerable stars whose light merges to give a milky effect when looking along the layer, in directions such as *D* and *E*.

pointing out that each of them was destined, after death, to pass either inwards or outwards. To force this home, he prepared a visual aid displaying a cross-section of the universe, and in it, by artistic licence, he portrayed the solar system and the visible stars as they actually appear from Earth; the light of the more distant of these stars, he said, merged to form 'a faint circle of light', the Milky Way.

It was only later than he realized his mistake: such a Milky Way would lie in every single cross-section of the universe that passed through both the Divine Centre and the solar system, whereas the real Milky Way is unique. In his beautifully illustrated *An Original Theory or New Hypothesis of the Universe*, published in 1750, he met this difficulty by greatly reducing the thickness of the spherical shell of space occupied by the Sun and the other stars of our system (for now he envisaged many such systems, each with its own Divine Centre). As a result, when looking inwards or outwards we see only a few, near (and therefore bright) stars, before our gaze extends into empty space. But when we look tangentially to the shell, whose radius is vast and which therefore curves imperceptibly, we see great numbers of stars whose light merges to create a milky effect: that is, the plane of the Milky Way is tangent to the shell at our particular location.

A summary of Wright's book, but without the illustrations necessary to comprehend his bizarre conceptions, appeared in a Hamburg periodical the following year and came to the attention of the German philosopher Immanuel Kant (1724–1804). Kant assumed, not unreasonably, that there must be only one Divine Centre, away in some remote part of the universe, and that the region of our star system was therefore entirely in the natural order. He knew of milky patches (nebulae) that had been observed in the sky and which he believed to be other star systems; but these were elliptical, whereas a spherical system seen from without will always appear circular. Kant therefore opted for an alternative model offered by Wright, in which the stars surrounding our Divine Centre formed a flattened ring. As far as Kant was concerned, there seemed no reason why this ring of stars (being wholly in the natural order) should not extend without a break from one side to the other, thus forming a complete disc of stars; and a disc seen edge-on will appear elliptical, just like the nebulae that had been observed. Kant therefore credited Wright – mistakenly – with the conception of the Milky Way as a disc-shaped aggregate of stars, which indeed it is.

These and similar speculations, dreamt up by amateurs in the warmth of the study, were hardly likely to impinge on professional astronomers. On the other hand they could scarcely ignore the discovery of the planet Uranus in 1781 by another amateur, the familiar William Herschel, a musician who had come to England from Hanover as a refugee from the Seven Years' War. But the amateurishness of his report, and his casual claim to have made the discovery with eyepieces whose alleged magnifications were beyond the powers even of professional opticians, made him a figure of controversy.

In 1772 Herschel had rescued his sister Caroline from family servitude in Hanover, and she was to be his devoted assistant in everything he did. His enthusiasm for astronomy soon began to take over their lives, Herschel's ambition being nothing less than to understand 'the construction of the heavens'. Herschel realized that to see objects that were distant and therefore faint, he must equip himself with reflecting telescopes able to collect as much light as possible – in other words, with the largest possible mirrors. He learned to grind and polish discs purchased from local foundries, but his ambitions soon outreached their capacities: a 3-foot disc he would have to cast himself. Undaunted, in 1781 he turned the basement of his own home into a foundry, but his two attempts resulted in failure and near-tragedy.

The discovery of Uranus gave Herschel's admirers the opportunity to lobby the King on his behalf, and in 1782 Herschel was awarded a royal pension that enabled him to devote himself to astronomy. He moved to the vicinity of Windsor Castle, where he had no duties

18. William Herschel's 'large' 20-foot reflector, commissioned in 1783, from an engraving published in 1794. His 'sweeps' with this instrument resulted in the discovery of 2,500 nebulae and clusters of stars. By 1820 the woodwork was in an advanced state of decay, and William's son John was forced to build a replacement, which he took to the Cape of Good Hope to extend his father's work to the southern skies.

except to show the heavens to the royal family and their guests when asked. He soon constructed one of the great telescopes of history, a reflector of 20-foot focal length and mirrors 18 inches in diameter and, equally important, a stable platform.

With Caroline seated at a desk within earshot, ready to act as amanuensis, Herschel devoted much of his observing time over the next two decades to 'sweeping' the night sky for nebulae and clusters of stars. The telescope, facing south, would be set to a given elevation, and alternately raised and lowered a little from this position as the heavens rotated slowly overhead; in this way a strip of sky would be 'swept' for any nebulae it might contain. When they began, only a hundred or so of these mysterious objects were known; when they finished, they had collected and classified 2,500 specimens.

Everyone recognized that a star cluster so distant that the individual stars could not be distinguished would appear nebulous, as indeed did the Milky Way. But were all nebulae distant clusters, or were some formed of nearby clouds of luminous fluid ('true nebulosity', as Herschel termed it)? If a nebula were seen visibly to alter shape, this would prove it was a nearby cloud, for a distant cluster would be too vast to change so rapidly; and in 1774, on the very first page of his first observing book, Herschel had noted that the great nebula in Orion was not as it had been depicted earlier (by Huygens in the 17th century). Occasional observations of the same nebula in subsequent years persuaded him that it was continuing to change, and was therefore formed of true nebulosity. But how to distinguish true nebulosity from a distant star cluster? It seemed to Herschel that he was encountering two kinds of nebulosity, milky and mottled, and he supposed that mottled nebulosity reflected the presence of innumerable stars.

But then, in 1785, he came across a nebula that contained individual stars together with both kinds of nebulosity; and he interpreted this as a star system that extended away from the observer. The nearest stars were individually visible, the more distant appeared as mottled nebulosity, while the most distant appeared as milky nebulosity. Herschel therefore reversed his earlier position, and decided that all nebulae were star clusters.

But clusters implied clustering: an attractive force or forces – presumbly Newtonian gravity – that was operating to pull the member stars together ever more closely. This implied that in the past, the stars of a cluster had been more scattered than they were now; in the future, they would be more tightly clustered.

In this way Herschel was introducing into astronomy concepts from biology: he was acting the natural historian, collecting and

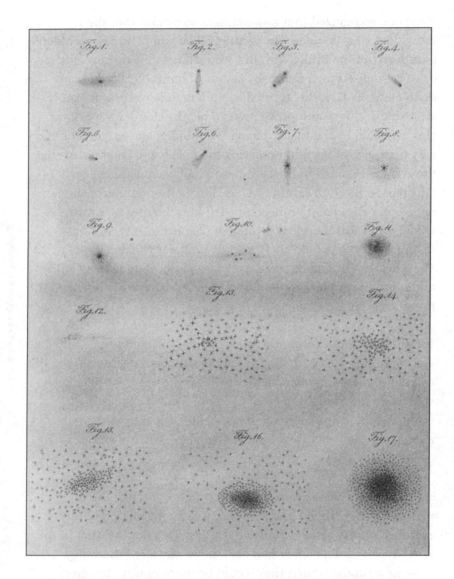

19. Sketches by Herschel showing objects from his catalogues of nebulae and clusters of stars, arranged in order of increasing maturity: gravitational attraction operates to concentrate the clusters more and more as time passes. From *Philosophical Transactions*, vol. 104 (1814).

classifying great numbers of specimens, which he could arrange by age – as young, in middle life, and old. He was changing the very nature of the science.

One evening in 1790, he was sweeping as usual when he came across a star surrounded by a halo of nebulosity. The star, he admitted, must be condensing out of the nebulosity, and so true nebulosity must exist after all. He would have to extend his theory of the development of star systems back in time, to accommodate an earlier phase during which thinly scattered light condensed under gravity into clouds of nebulosity, out of which stars were born. These stars formed clusters, scattered at first but increasingly condensed – until at length the clusters collapsed upon themselves in a great celestial explosion, the light from which began the cycle over again. Herschel's contemporaries, almost none of whom had instruments with which to view the evidence, were at a loss to know what to make of it all.

It was William Herschel's son John (1792–1871) who brought stellar astronomy into the scientific mainstream. When he was a young man his father prevailed upon him to abandon a career at Cambridge and return home, to become in effect his apprentice and astronomical heir, charged with the re-examination and extension of his father's collections of astronomical specimens. William's 20-foot reflector was now decayed with age, but before he died in 1822, he supervised John in the construction of a replacement.

In 1825 John Herschel began the revision of his father's catalogues of nebulae visible from England. This done – and resolutely declining all offers of government help – he set sail for the Cape of Good Hope, where he spent four years extending to the southern skies his father's catalogues of nebulae, double stars, and so forth. He became, and remains, the only observer to have examined the entire celestial sphere with a major telescope.

When John Herschel set sail for home in March 1838, his career as an observer was ended, and so was the Herschelian monopoly on great telescopes. That year, at Birr Castle in central Ireland, William Parsons (1800–67), the future Earl of Rosse, manufactured and assembled the segments to form a composite mirror 3 feet in diameter. The following year he succeeded in casting a single disc of this size, and in 1845 he completed the 'Leviathan of Parsonstown', a monster reflector slung between huge masonry walls, with mirrors no less than 6 feet in diameter and weighing four tons each. Within weeks the reflector had revealed that some nebulae are spiral in structure.

The Leviathan was designed to settle once and for all the question of whether all nebulae were merely star clusters disguised by distance; and all agreed that examination of the great nebula in Orion, which is visible to the naked eye, would be decisive. It so

**20. Lord Rosse's reflector with 6-foot mirrors, commissioned in 1845. By April of that year, Rosse had used it to discover the spiral structure of some nebulae.**

happened that the reflector was powerful enough to detect stars that are indeed embedded in this (gaseous) nebula, and on seeing these Rosse persuaded himself that he was viewing a cluster that he had triumphantly 'resolved' into its component stars.

Many agreed that this success with the greatest of the nebulae could be generalized, and that belief in 'true nebulosity' had been banished from astronomy for ever. They were soon to be proved wrong, but only after astronomy had surrendered its autonomy and merged with physics and chemistry to embark on the analysis of starlight.

# Epilogue

In this book we have followed attempts of observers and theoreticians down the ages to understand the heavenly bodies – what they are, and how they behave. Whether observers consciously thought of it or not, their information came from the light currently arriving at Earth from these bodies: it was light, rather than the bodies themselves, that they observed.

Not all such light was the same. Some stars shone with a brilliant whiteness, for example, while others had a reddish hue. The relationship between colours and the white light that arrives from our nearest star, the Sun, was established in 1666 by Isaac Newton. He made a hole in the shutters of his room in Trinity College, Cambridge, and passed the beam of sunlight through a prism. He saw, as expected, the familiar spectrum with all the colours of the rainbow. The received theory was that white light was simple and basic, and that a colour resulted from some sort of modification of white light: you started with white light, and you did something to it to obtain a colour. By careful experimentation Newton found that, to the contrary, it was the colours that were basic, and that when recombined together they formed white light once more. Sunlight was made up of the colours of the rainbow.

Newton was investigating light itself, and not the Sun as source of the light. William Herschel was the first observer with the curiosity to examine the spectra of light from other stars, and with telescopes of sufficient 'light-gathering power' to make this a practical possibility. As early as 1783 he several times put a prism at the eyepiece of one or other of his 20-foot reflectors when it was directed to a bright star, but it was not until 9 April 1798 that he undertook a brief investigation into the light from six of the brightest stars. 'The light of Sirius', he found, 'consists of red, orange, yellow, green, blue, purple, and violet'. On the other hand, 'Arcturus contains more red and orange and less yellow in proportion than Sirius'. And so on. But what these differences implied he had no idea.

The answer was gradually to emerge from a much more careful analysis of sunlight. In 1802 William Hyde Wollaston (1766–1828) repeated Newton's experiment, but he replaced the crude hole that Newton had made in his shutters with a narrow slit only one-twentieth of an inch wide. He was surprised to find the spectrum of sunlight was crossed by seven dark lines, which he took to be the divisions between the colours. However, when the telescope-maker Joseph Fraunhofer was making tests on glass lenses, he was astonished to discover that there were in fact hundreds of such lines. He also found that spectra that were quite different in form could be generated in the laboratory, consisting of thin bright lines with dark spaces between them (a 'bright-line' spectrum, in contrast to the continuous spectra of the Sun and stars).

Over the next three decades, the situation was gradually clarified, and the far-reaching implications became apparent. Two Germans, the chemist Wilhelm Bunsen (1811–99) and the physicist Gustav Robert Kirchhoff (1824–87), played a central role in this. By 1859 they had established that glowing solids and liquids produced a continuous spectrum, that of sunlight being the familiar example, while glowing gases produced a bright-line spectrum. (As a result, when the English astronomer William Huggins (1824–1910) in 1864 managed to obtain a visible, bright-line spectrum from a nebula in Draco, he ended the centuries-old debate as to whether 'true' (gaseous) nebulae exist.) Each element had its own characteristic line positions. Surprisingly, a continuous spectrum when passed through a gas displayed a 'dark-line' spectrum, with dark lines characteristic of the gas. As a result, once the line positions of an element had been established in the laboratory, the investigator could demonstrate the presence or absence of the element in the star or nebula, or in any gas through which the light from the celestial body had passed.

A Pandora's box was opened. As the great American observer James Keeler remarked, 'The light which reveals to us the existence of the heavenly bodies also bears the secret of their constitution and physical condition'. The limitation on human knowledge famously declared in 1835 by Auguste Comte, that we can never by any means investigate the chemical composition of celestial bodies, was sensationally disproved. So profound was the transformation that astronomy lost its identity, to become a branch of physics (and chemistry). As Huggins put it,

> Then it was that an astronomical observatory began, for the first time, to take on the appearance of a laboratory. Primary batteries, giving forth noxious gases, were arranged outside one of the windows; a large induction coil stod mounted on a stand on wheels so as to follow the positions of the eye-end of the telescope, together with a battery of Leiden jars; shelves with Bunsen burners, vacuum tubes, and bottles of chemicals, especially of specimens of pure metals, lined its walls.

The study of the properties, constitution, and evolution of heavenly bodies became the province of 'astrophysicists', rather than astronomers, and Kepler's title, 'the new astronomy', was invoked once more. Meanwhile, traditional astronomy flourished and developed, in tandem with astrophysics.

The history of any science is never-ending, and the scope of the discipline grows while the number of practitioners escalates. The transformation of astronomy in the mid-19th century marks the end of the story we have set out to tell, and – as always – the beginning of another.

Astronomical research is now the work of teams of scientists and engineers. Radio telescopes intercept radiation – information – at wavelengths imperceptible to the human eye, and combinations of such telescopes may be equivalent to a single 'dish' hundreds of kilometres in diameter. Optical telescopes with ever-larger mirrors are being built on the tops of mountains above most of the Earth's atmosphere, and predominantly in the southern hemisphere where many of the most significant deep-sky objects are to be seen. Computers drive the telescopes, and by 'active optics' continuously compensate for subtle changes in the balance of the mirror and in the atmosphere above the instrument. New technology has enormously increased the amount of incoming information that can be secured from some of the most dramatic of contemporary instrumentation, including the Hubble Space Telescope and the planetary probes. Viewing the images these spacecraft transmit to Earth by radio links makes it easy to understand why there has never been a more exciting time to be an astronomer.

# Further reading

The present work is in effect an introduction to two closely related books by the author and colleagues, either of which would serve as a text for further reading. They are: Michael Hoskin (ed.), *The Cambridge Illustrated History of Astronomy* (hereafter *CIHA*; Cambridge, 1997); and Michael Hoskin (ed.), *The Cambridge Concise History of Astronomy* (*CCHA*; Cambridge, 1999). The *Illustrated History* has numerous illustrations in colour, while in the *Concise History* the text (despite the book's title) is amplified with additional technical material. The subjects of our first four chapters are further discussed in articles in Christopher Walker (ed.), *Astronomy Before the Telescope* (*ABT*; London, 1996). The relevant sections of these works are given first in the suggested further reading below. For an alternative overview of the whole history of astronomy, see the paperback by John North, *The Fontana History of Astronomy and Cosmology* (London, 1994).

All the books cited above include bibliographies. Individual astronomers are treated authoritatively in the multi–volume *Dictionary of Scientific Biography*, edited by C. C. Gillispie (New York, 1970–90), available in many reference libraries.

Those wishing to keep abreast of current work in the field may consult the *Journal for the History of Astronomy* (Science History Publications, Cambridge).

## Chapter 1

*CIHA* or *CCHA*, Chapter 1; *ABT*, article by Ruggles.

The customs of orienting buildings on heavenly bodies in prehistoric Europe and the Mediterranean area are discussed in Michael Hoskin, *Tombs, Temples and Their Orientations: A New Perspective on Mediterranean Prehistory* (Bognor Regis, 2001). For the British Isles, see Clive Ruggles, *Astronomy in Prehistoric Britain and Ireland* (New Haven and London, 1999), a more technical work with discussions of methodology.

## Chapter 2

*CIHA* or *CCHA*, Chapter 2; *ABT*, articles by Wells, Britton and Walker, Toomer, Jones, and Pingree.

A wide-ranging and user-friendly book is James Evans, *The History & Practice of Ancient Astronomy* (New York and Oxford, 1998). Otto Neugebauer, *Exact Sciences in Antiquity*, 2nd edn. (Providence, RI, 1957), is somewhat dated, but the work of a master. Astrology was a powerful motivation for astronomy in antiquity; the best account is Tamsyn Barton, *Ancient Astrology* (London and New York, 1994).

## Chapter 3

*CIHA* or *CCHA*, Chapters 3 and 4; *ABT*, articles by Field, King, and Pedersen.

On astronomy in Christendom, see Stephen C. McCluskey, *Astronomers and Cultures in Early Medieval Europe* (Cambridge, 1998), and Edward Grant, *Planets, Stars, and Orbs: The Medieval Cosmos, 1200–1687* (Cambridge, 1994).

## Chapter 4

*CIHA* or *CCHA*, Chapter 5; *ABT*, articles by Swerdlow and Turner.

Astronomy of the period is treated systematically in *The General History of Astronomy*, Vol. 2: *Planetary Astronomy from the*

*Renaissance to the Rise of Astrophysics*, edited by R. Taton and
C. Wilson, Part A: *Tycho Brahe to Newton* (Cambridge, 1989).

## Chapter 5
*CIHA* or *CCHA*, Chapter 6.

The General History of Astronomy, Vol. 2, Part A, Chap. 13 is the best
introduction to Newton's *Principia*, while Part B of the same work, *The
Eighteenth and Nineteenth Centuries* (Cambridge, 1995) is excellent on
the implementation of the Newtonian programme.

## Chapter 6
*CIHA* or *CCHA*, Chapter 7.

Michael Hoskin, *Stellar Astronomy: Historical Studies* (Cambridge,
1982: Science History Publications, 16 Rutherford Road, Cambridge
CB2 2HH).

# Glossary

**aberration of light**: the small, constantly varying shift in the observed position of a star caused by the velocity of the Earth-based observer's orbit around the Sun.

**annual parallax**: the small, constantly varying displacement in the observed position of a star, caused by the displacement of the Earth-based observer from the centre of the solar system.

**astronomical unit**: the mean distance of the Earth from the Sun (about 93 million miles).

**atmospheric refraction**: the bending of the path of light from a celestial body by the Earth's atmosphere.

**binary star**: two stars that are companions in space, bound to each other by their mutual attraction and each in orbit about their common centre of gravity.

**bright-line spectrum**: see *spectrum* below.

**celestial equator**: the projection onto the sky of the equator on Earth.

**celestial poles**: the projection onto the sky of the axis of rotation of the Earth.

**centrifugal force**: the tendency of a body in orbit to 'fly off at a tangent'.

**centripetal force**: a force such as gravity that causes a body to move 'inwards' in a curved path, instead of continuing in a straight line.

**dark line spectrum**: see *spectrum* below.

**deferent**: in ancient or medieval planetary theory, the circle in the geometric model of a planet's orbit about the Earth that 'carries' either the Sun or the centre of the little circle or 'epicycle' on which the planet is imagined as located.

**eccentric circle**: in ancient or medieval planetary theory, a circle that is not centred on the Earth.

**ellipse**: a closed curve formed by the intersection of a cone with a plane; as Kepler showed, the planets orbit the Sun in elliptical paths.

**empty focus**: as a planet orbits the Sun in its elliptical path, the Sun is located at one of the two foci (a geometrically significant position on the major axis of the ellipse, to one side of the centre); the 'empty' focus is the matching position on the other side of the centre, and is so called because no physical body is located there.

**epicycle**: in ancient or medieval planetary theory, the small circle that carries the planet in the geometrical model of a planet's orbit about the Earth, and which is itself carried on the deferent.

**equant point**: in the geometrical model of a planet's orbit in Ptolemy's *Almagest*, the position symmetrically opposite the eccentric Earth; the planet is imagined moving with a (variable) speed such that, viewed from the equant point, its motion across the sky appears uniform.

**heliacal rising**: the reappearance in the dawn sky of a star or planet after several weeks of invisibility lost in the glare of the Sun.

**inverse-square force**: a force, such as gravity, that reduces as distance increases, in proportion to the square of the distance.

**nebula**: a milky patch seen in the sky and so quite different in appearance from a star or planet; physically, nebulae are of several different kinds, some being vast star systems so far away that it is difficult to distinguish the individual stars, while others are wholly or partly gaseous.

**nova**: *nova stella*, a 'new' star that appears where none was visible before.

**Olbers's Paradox**: if the stars were scattered at regular intervals throughout infinite space, analysis suggests that the whole of the sky would appear as bright as the Sun; modern cosmologists have erroneously supposed that the fact that this does not happen was seen as a paradox by H. W. M. Olbers (1758–1840).

**period (of a planet)**: the time taken by a planet to complete one orbit of the Sun.

**precession of the equinoxes**: the slow change of direction of the Earth's axis of rotation (period 25,800 years), as a result of which the celestial equator moves relative to the ecliptic (the path of the Sun), and their points of intersection—the 'equinoxes'—therefore move or 'precess'.

**proper motion (of stars)**: the observed motion of an individual star on the celestial sphere (and so at right angles to the observer's line of sight).

**rectilinear inertia**: the tendency of a moving body to continue moving in a straight line with uniform speed.

**retrograde**: the occasional 'backwards' motion of a planet as seen from Earth; for example, the motion of Mars, Jupiter or Saturn appears retrograde for a time when the Earth overtakes it 'on the inside', as the planets together orbit the Sun.

**spectrum**: the continuous band of 'colours of the rainbow', from violet to red, into which white light is dispersed when passed through a prism. White light is emitted by very hot solids, liquids, or dense gases. A *dark line spectrum* is crossed by transverse dark lines which are formed by cooler gases in the line of sight, e.g. in the cooler atmosphere of the Sun. A *bright line spectrum* has luminous lines on a fainter or dark background, which arise from a hot but highly rarified gas, e.g. in a gaseous nebula.

**variable star**: a star whose apparent brightness varies, either regularly or irregularly.

# "牛津通识读本"已出书目

古典哲学的趣味
人生的意义
文学理论入门
大众经济学
历史之源
设计，无处不在
生活中的心理学
政治的历史与边界
哲学的思与惑
资本主义
美国总统制
海德格尔
我们时代的伦理学
卡夫卡是谁
考古学的过去与未来
天文学简史
社会学的意识
康德
尼采
亚里士多德的世界
西方艺术新论
全球化面面观
简明逻辑学
法哲学：价值与事实
政治哲学与幸福根基
选择理论
后殖民主义与世界格局

福柯
缤纷的语言学
达达和超现实主义
佛学概论
维特根斯坦与哲学
科学哲学
印度哲学祛魅
克尔凯郭尔
科学革命
广告
数学
叔本华
笛卡尔
基督教神学
犹太人与犹太教
现代日本
罗兰·巴特
马基雅维里
全球经济史
进化
性存在
量子理论
牛顿新传
国际移民
哈贝马斯
医学伦理
黑格尔

地球
记忆
法律
中国文学
托克维尔
休谟
分子
法国大革命
民族主义
科幻作品
罗素
美国政党与选举
美国最高法院
纪录片
大萧条与罗斯福新政
领导力
无神论
罗马共和国
美国国会
民主
英格兰文学
现代主义
网络
自闭症
德里达
浪漫主义
批判理论

德国文学

戏剧

腐败

医事法

癌症

植物

法语文学

微观经济学

湖泊

拜占庭

司法心理学

发展

农业

特洛伊战争

巴比伦尼亚

河流

战争与技术

品牌学

儿童心理学

时装

现代拉丁美洲文学

卢梭

隐私

电影音乐

抑郁症

传染病

希腊化时代

知识

环境伦理学

美国革命

元素周期表

人口学

社会心理学

动物

项目管理

美学

电影

俄罗斯文学

古典文学

大数据

洛克

幸福

免疫系统

银行学

景观设计学

神圣罗马帝国

大流行病

亚历山大大帝

气候

第二次世界大战

中世纪

工业革命

传记